TROPICAL MARINE BIOLOGY

TROPICAL MARINE BIOLOGY

Danielle May
Joan Aquino

Kruger Brentt
Publishers

2024

Kruger Brentt Publishers UK. LTD.
Company Number 9728962

Regd. Office: 68 St Margarets Road, Edgware, Middlesex HA8 9UU

ISBN: 978-1-78715-192-5

For information on all our publications visit our website at http://krugerbrentt.com/

PREFACE

Tropical marine biology is the study of the marine life and ecosystems in tropical regions of the world. These regions are known for their high biodiversity and unique species that have evolved to survive in the warm and nutrient-rich waters. In this essay, we will explore the importance of tropical marine biology, the challenges it faces, and the advancements made in the field. The importance of tropical marine biology lies in the fact that these regions support some of the most diverse and complex ecosystems on the planet. Coral reefs, for example, are found primarily in tropical regions and provide habitat and food for millions of species. These reefs also serve as natural barriers that protect coastlines from storm surges and erosion. In addition, tropical marine ecosystems are a source of food, medicine, and other valuable resources that are critical to human livelihoods. However, tropical marine biology faces several challenges that threaten the health of these ecosystems. Climate change, for example, is causing rising sea temperatures and ocean acidification, which can have detrimental effects on coral reefs and other marine life. Overfishing, pollution, and habitat destruction are also major threats that can lead to the decline of species and the degradation of marine ecosystems.

Despite these challenges, there have been significant advancements in tropical marine biology in recent years. Scientists are using new technologies, such as underwater drones and DNA sequencing, to study marine life and ecosystems more efficiently and accurately. Researchers are also developing innovative approaches to conservation and restoration, such as coral reef restoration programs that use artificial reefs to rebuild damaged ecosystems. Tropical marine biology plays a crucial role in understanding and protecting the diverse and complex ecosystems of tropical regions. Despite facing significant challenges, there have been advancements in the field that hold promise for the future of marine conservation.

The present book describes marine life and marine processes that ranges from the unimaginably abundant microscopic organisms that drive the oceans' food web to the apex predators. It provides fundamental overview of the tropical marine biology. The subject matter is presented in a lucid and succinct style so that even those who do not have a prerequisite knowledge of the subject can easily understand it. This book is perfect for anyone who wants to know more about the life under the waves, amateur naturalists, conservationists and marine biologist.

We are grateful to all those persons as well as various books, manuals, periodicals, magazines, journals etc. that helped in the preparation of this book. In spite of the best efforts, it is possible that some errors may have occurred into the compilation and editing of the book. Further queries, constructive suggestions and criticisms for the improvement of the book are always welcomed and shall be thankfully acknowledged.

Danielle May

Joan Aquino

CONTENTS

1 Chapter

WHY DO THE OCEANS MATTER?

Recently many people have been making a huge fuss about cleaning up our oceans and climate change. They talk about all of these different things: ocean warming, ocean acidification, overfishing, pollution, and coral bleaching. They tell us the oceans are in grave danger. So, yeah, all of this sounds really bad but honestly what does it all mean? More importantly, why should it matter to you?

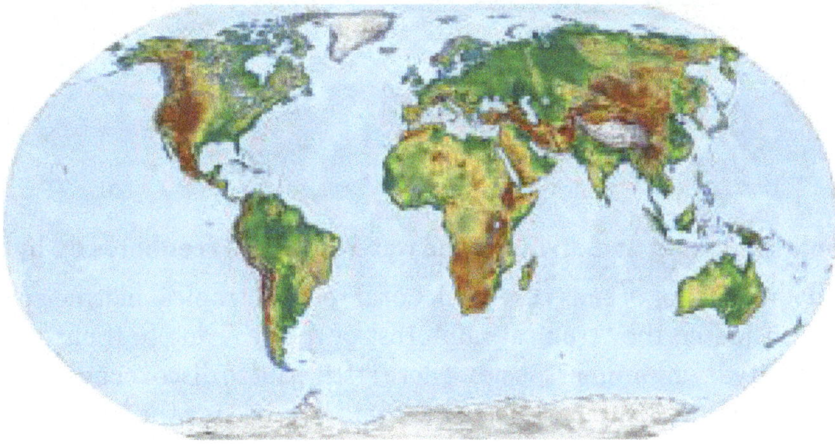

Physical Map of the World under CC 4.0

Looking at a map you can tell that the ocean takes up the majority of our planets surface area, and when calculated it covers a whopping 72% of the earths surface! The ocean also provides over 50% of the earth's oxygen and stores 50 times more carbon dioxide than the atmosphere. It also plays a huge role in regulating our planet's climate and weather patterns. If you want to read a little more about all of this check this article out.

If all of that doesn't convince you of the importance of the ocean, thin about all the fun that people can enjoy involving the ocean and the beach. There's swimming, surfing, boating, parasailing, cruises, the list goes on and on! Okay, so maybe you're not a beach person, I mean the sand does get everywhere.

The ocean also plays a pretty big part of the economies of the world. You also may not realize how important the ocean is for coastal communities in developing countries. For example just in the United States the ocean produces about 282 million dollars per year and creates about 3 million jobs. It also provides a mode for 90% of all trade transportation in the world. Lastly, we use the ocean for a large portion of our protein source.

People swimming and diving in the water by GoodFreePhotos CC by 2.0

The biodiversity of oceans is critical. Coral reefs (a tropical marine ecosystem) are often considered the "tropical rainforest of the sea" due to their diversity in marine life and varying animal species. There is still a lot to discover about the ocean, however we do know it's important to maintain because it's the largest ecosystem we have on the planet. This ecosystem maintains homeostasis via regulation by abiotic and biotic factors. The oceans are often overlooked especially since we don't see the dynamics between marine organisms and our own environment as we don't share this habitat.

In tropical marine ecosystems, biotic factors include sea plants, algae, bacteria, and animals. Abiotic factors are similar to those on land such as sunlight, temperature, and soil nutrients. However, since the ocean is water, pH and salinity are also considered important abiotic factors that influence and sustain healthy and productive organisms. Oceans function just like land ecosystems, but they have far more area and depth. Both aquatic and terrestrial ecosystems have complicated food webs that distribute energy, and the main biotic components are a variety of primary producers, consumers and decomposers. Marine food webs contain primary producers such as phytoplankton which undergo photosynthesis and provide consumers with the nutrients they need. Click here for a cool video from Khan Academy to learn more about about food chains and trophic levels. Trophic pyramids for aquatic ecosystems are inverted compared to terrestrial ecosystems due to the fact that the short life span and rapid consumption of autotrophic phytoplankton means the biomass of heterotrophs at any given time might be greater than autotrophs (the opposite of what you usually see in terrestrial systems).

Although the ocean is somewhat overlooked it is a truly remarkable ecosystem. Unfortunately, there has been a rise in ecological crisis in aquatic ecosystems mostly at the hands of humans. For example, overfishing can alter predator/prey balance, and pollution can significantly impact the health of the environment. If the oceans continue to be in crisis at a global level, it could result in mass species extinction. Therefore, it is really important to care for our oceans. We hope this chapter gives you a little more perspective on why the ocean is so important for a healthy environment, global economy and for civilizations around the world.

2 Chapter

CORAL REEFS: AN INTRODUCTION

What are corals you ask? " A little bit algae, a little bit rock, and a lot animal." Corals are close cousins to sea anemones and jellyfish, meaning they all belong to the same phylum of Cnidarians. Coral is a marine invertebrate that is composed of simple structures known as polyps. The polyp has one opening which is the coral's mouth that is surrounded by a ring of tentacles. These tentacles have stinging cells called cnidocytes, used for food and protection. These cells fire tiny toxic barbs called nematocysts when they are touched to ward off predators or capture food. Inside the polyp are reproductive and digestive tissues. Coral can live as an individual polyp, but most corals exist as large colonies of interconnected polyps.

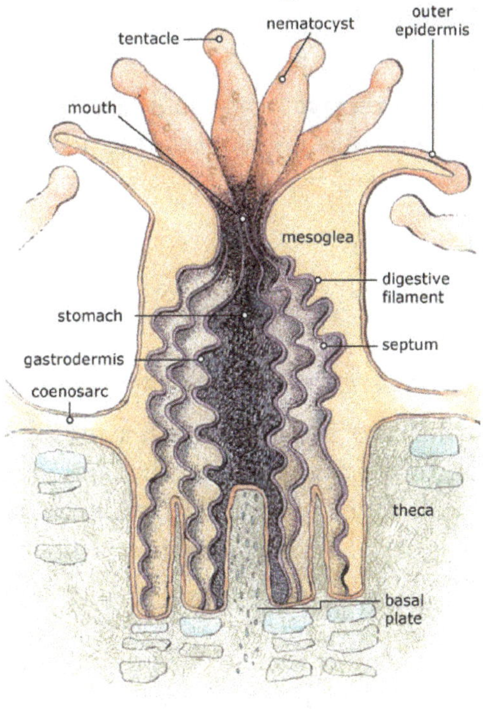

"Coral Polyp" by Wikimedia Commons under CC by 2.0

HOW DO CORALS SURVIVE?

- **Sunlight:** Corals grow in shallow water where the sunlight can reach them. The algae that live inside of them, zooxanthellae, need sunlight to survive, since the coral animal depends on the zooxanthellae, corals need sunlight to survive.
- **Clear Water:** Clear water is needed for corals to survive because it lets the sunlight in. When the water is opaque the corals will not thrive.
- **Warm Water Temperature:** Most corals, though not all, require warm waters to grow and survive. Corals general live in water temperature of 68-90°F
- **Clean Water:** Corals are very sensitive to pollution and sediments that can be in the sea. Sediments can create cloudy water conditions which won't allow for sunlight to get through which will harm the polyps. Corals do not typically do well in environments where there is little sunlight, except for the deep-sea corals. Another issue that may arise is the addition of too many nutrients. An overflow of nutrients in the water can cause things such as algae to grow, in turn taking over the reef. Sediments and pollutants can also cause direct harm to coral tissue.
- **Salt Water:** Corals require a certain balance between the concentration of salts and water (salinity).
- **Nutrients:** Though zooxanthellae do provide some nutrients, the coral animals still need to receive other sources of nutrition. Corals can also grab passing nutrients (like zooplankton) by using their tentacles to snag passing organisms flowing by.

CORAL REEF FUNCTIONS

- **Protection:** Functions as a form of protection for the shores from harsh waves and storms
- **Habitat:** Provides shelter and safety for many organisms
- **Nutrients:** Many forms of heterotrophic marine life are supported from the primary production of the coral zooxanthellae. Carbon and nitrogen-fixing is an important part of the marine food chain. Nutrient cycling within the coral polyps is very efficient and contributes to the high production of reefs.

HOW DO CORALS REPRODUCE?

Some species have distinct male or female polyps while other coral can alternate gender. Many coral reef species only reproduce once or twice a year. For some coral that only have single-gender polyps, certain events (i.e. full moon) trigger the polyps to release massive amounts of sperm or eggs that flood into the seas, where the gametes eventually come together. Most coral species spawn by releasing eggs and sperm into the water; but the period of spawning depends on the type of coral. When an egg and sperm meet, the fertilized egg eventually develops into a larva called planula. Planula can form by two different ways, either fertilized within the body of a polyp or fertilized outside the polyp, externally in the water. Fertilization of an egg within the body of the coral polyp happens when sperm that is released through the mouth of another polyp, is taken in by a polyp with an egg. When the larva is matured enough the mother will spit it out via the mouth. If fertilization is external, corals will eject large amounts of egg and sperm into the water which eventually find each other. The release of larvae or gametes is known as coral spawning. Most species also reproduce asexually, budding new polyps or fragments, that grow either on an older polyp or drift around before landing on a new surface.

"Cauliflower coral spawning, French Frigate Shoals" by USFWS – Pacific Region under CC by 2.0

The information in this chapter is thanks to content contributions from Haley Zanga, Marisa Benjamin, and Haley Fantasia

3 Chapter

DIFFERENT TYPES OF CORALS

WHAT TYPES OF CORAL EXIST?

When people think about coral they often think about the stony calcified coral they get at a gift shop during a vacation to a tropical island. However, these are only the skeletons of stony corals, and are only one type of coral that exists. There are also deep water and soft corals. Each of these types of corals is unique structurally and in how they interact with their surroundings. There are over two thousand different types of coral which form colonies that play a key role in marine environments.

1. HARD CORALS

"Hard Corals" by Amal FM under CC BY-SA 2.0.

Hard coral is a fundamental part of building the coral reef. The polyps of hard corals secrete skeletons of calcium carbonate (Limestone), that will eventually become rock. Hard coral includes species such as brain coral and elkhorn coral. Hard coral are considered to be hermatypes, which are reef-building corals. These corals require a certain type of algae called zooxanthellae that live within their tissues, this important mutualism is necessary for their survival. These organisms are what give corals their colors, which is dependent upon where they are living. The shape and appearance of each coral are dependent upon its species type, location, depth, water movement, and many other factors. Hard coral polyps are very similar to sea anemones, as they also have stinging cells called cnidocytes. Hard coral species can be found in all of the world's oceans, yet their populations are expected to decrease due to global changes and ocean acidification.

As mentioned earlier, hard corals build coral reefs. A coral reef begins when the coral polyp attaches itself to a rock on the seafloor. This one coral polyp then begins to divide or bud into thousands and thousands of clones of itself. Polyps and their calcareous skeletons connect to one another which create a colony that will act as one. As this colony grows over hundreds and thousands of years it will join with other colonies and become a reef. Read more in our chapter on coral reef formation.

"Maldives reef" by Alexander Semenov under CC by 2.0

2. SOFT CORALS

"Soft Corals" By Najim Almisbah Under Cc By-Sa 2.0

Soft corals often resemble plants and trees. They are soft and bendable so they do not have stony skeletons like hard coral. For protection and support, they grow woodlike cores for stability. These cores are made of structural proteins such as gorgonin and other protein similar to those of nails and horns of other animals. These types of coral are referred to as Ahermatypes (non-reef building corals). Examples of soft coral in the Bahamas and Caribbean include sea fingers or sea whips. This type of coral also does not always have a symbiotic relationship with zooxanthellae. Though many utilize their presence, soft corals will typically eat any type of passerby out of the water column. These types of corals do not produce calcium carbonate, rather they contain spiny skeletal sclerites. These give the softer corals some protection and support. Soft species also prefer to live in nutrient-rich waters with less intense light. In other parts of the world, fleshy true soft corals have no rigid internal skeleton at all.

3. DEEP-SEA CORALS

Deep Sea Coral, White "Black Coral" ***Leiopathes Glaberrima*** **By Noaa Photo Library Under Cc By 2.0**

Deep-sea coral can be found in the dark depths of up to 6,000 m (20,000 ft) below the surface of the ocean. These corals live in icy cold water with little to no sunlight. These corals, like shallow species, can exist as single polyps or multiple, living in complex colonies made up of different species. Since these species do not require sunlight or warm water, they are able to grow in a vast array of waters around the world. They have even been found in waters as cold as -1-degree celsius. Since these corals live without sunlight they contain no zooxanthellae. This means these corals must obtain their energy and nutrients elsewhere. They do this by trapping tiny organisms in passing currents. The main reason many scientists had no idea of the existence of these deep-sea corals, is because, for many years, the oceans deep depths were inaccessible. In addition to being surprisingly diverse, scientists have also discovered that deep-sea corals are amongst the oldest marine organisms on record. Since corals are constantly growing, regenerating new polyps, some coral reefs have been actively growing for almost 40,000 years. To your and my surprise, scientists have identified nearly as many deep-sea corals as shallow-water species.

The information in this chapter is thanks to content contributions from Haley Zanga, Marisa Benjamin, Haley Fantasia, and Melissa Wydra

4 Chapter

REEF TYPES AND HOW CORAL REEFS ARE FORMED

Have you ever wondered how coral reefs are formed? It has been established by the scientists in the Tropical Marine Biology field that coral reefs can be divided into three types: Fringing reefs, Barrier reefs, and Atolls which all require very specific conditions to develop. Coral reefs begin to form when free-swimming coral larvae attach to a rock or other submerged object on the edges of islands. As the corals grow, over very long periods of time, they form into a reef. The three types of reef represent stages in development of a coral reef over time.

1. FRINGING REEFS

Fringing reefs grow near the coastline around islands and continents. These types of reefs are the most common type we see and are considered to be the youngest of the 3 types of coral reefs.

Fringing Reef in Eilat, Israel Poto by: Mark A. Wilson [CC by 2.0]

They are separated from the shore by shallow lagoons. The first stage of the formation is when the coral larvae attach itself to rocks or soil near the coasts. In certain parts of the world, they tend to form where volcanoes have also formed due to the shallow sloped walls being ideal to make shore reefs. The larvae become polyps and excrete calcium carbonate, which forms their exoskeleton. The secreted calcium carbonate sediments on the rocks and provides a substrate for more polyps coming to attach themselves. As more and more polyps attach, and layer over time, they create a coral reef. Calcareous Algae also add their sediments to the structure. Other organisms with calcareous skeletons, also add their own remains to the reef as they die and sink. Since fringing reefs are the youngest and grow onshore, they tend to have less diversity of species within that ecosystem. Some locations of these reefs include Kenya, Australia and other parts of Africa shores.

2. BARRIER REEFS

As the name states, these reef types border along coastlines with a very wide and deep lagoon separating the two structures. The term "barrier" comes from the more shallower coral that sticks out of the water and creates a barrier or wall-like structure. One of the most famous and largest barrier reefs is "The Great Barrier Reef" in Australia.

Great Barrier Reef by Studio Sarah Lou by Flickr

This reef is around 1200 miles and consists of many complex reefs making it up. This type of reef is formed when the fringing reefs slowly combine with each other and form a borderline along the coast. The calcium carbonate structures attract more polyps and the spaces are filled up. It forms a line along the coast and a ring around an island. Large barrier reefs are the rarest type of coral reef and only seen in a select few places on earth such as Belize and parts of the southern Pacific Ocean. Smaller barrier reefs are seen in places where islands are in an earlier stage of becoming submerged.

Most of the reefs surrounding the Turks and Caicos Islands are barrier reefs that formed from the precipitation of $CaCO_3$ on the remnants of continental debris resulting from the separation of North America from South Africa 200 million years ago (continental drift). Dissolved minerals in the seawater form small particles called ooids that become cemented together to form oolite rock which makes up most of the Turks and Caicos islands and bank.

3. ATOLL REEFS

An Atoll Reef is a ring-like shaped coral reef or small islands of reefs in a circle with a lagoon in the middle and are usually located in the middle of the sea. These types of reefs are formed when an island has sunk completely (or nearly) in the middle of the ocean from rising sea levels around a pre-existing structure (these islands are often the tops of underwater volcanoes). Most atoll reefs can be found in the Pacific ocean where the waters are warm and salty. They can grow between the sizes of 12.6 km^2 (Great Chagos Bank) and as small as 1.5 km^2 (Ontong Java reef). More animals are found in an atoll than in earlier stage reefs. Over a million species can live in one square kilometer of coral reef. The three types of reef from fringing to barrier to atoll, form in this sequential order and represent stages of development of a reef.

"Formation of Coral" by NOAA Ocean Service Education via Wikimedia Commons

Below is a link to a video created by three undergraduates at Keene State College who discuss more about the different types of coral reefs:

https://www.youtube.com/watch?v=3-0BgmkLGPI

4. PATCH REEFS

Patch reefs are small, isolated reefs that grow from the open bottom of the island platform or continental shelf. They usually occur between fringing reefs and barrier reefs. They vary in size and rarely reach the surface of the water.

Patch reefs in Hoopers Bay area by James St. John Via Flickr

The information in this chapter is thanks to content contributions from Haley Zanga and Emily Michaeles

5 Chapter

CORAL REEF COLORS

When you first see this picture below of a coral reef landscape, you are captivated. Captivated by the beautiful, bright, elegant colors popping through the clear blue water. But why? Why do these coral reefs have such bright, bold colors? The colors found in colorful corals are mostly due to three things – photosynthetic pigments, fluorescent proteins, and non-fluorescent chromoproteins.

"Colorful underwater landscape of a coral reef" by Wikimedia Commons under Public Domain

CORAL POLYPS

In order to understand why corals get their color, it is important to first learn the structure of the coral polyps. The majority of polyps have clear, transparent bodies over their hard, white skeletons. Millions of zooxanthellae live inside the tissues of these polyps. These zooxanthellae produce pigment, and because they reside in the clear tissue of the polyp, the pigments are visible, and the corals get their beautiful colors.

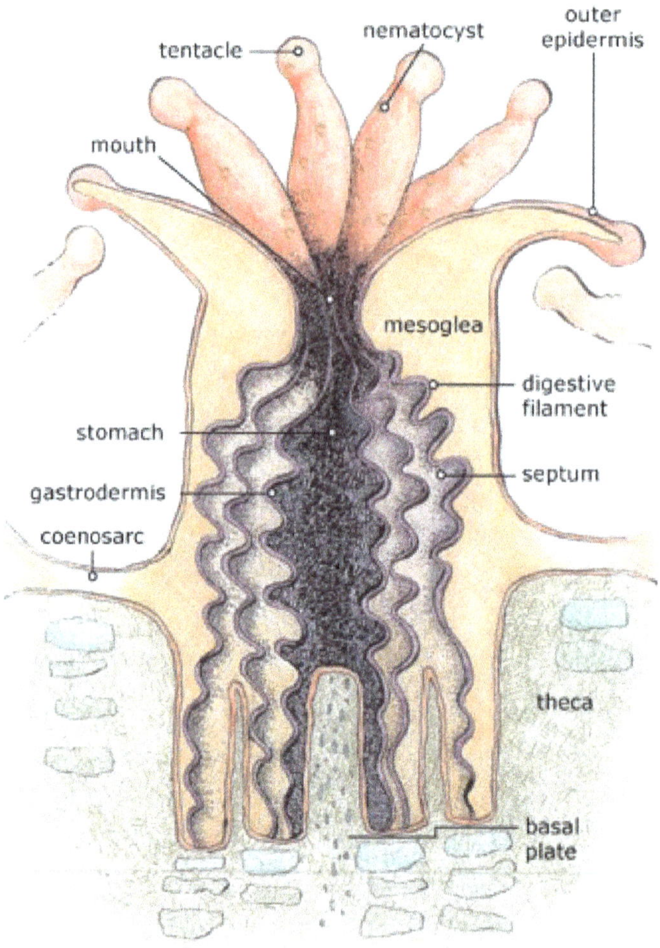

"Anatomy of a coral polyp" by Wikimedia Commons under public domain

Light Intensity Dependency

Zooxanthellae are photosynthetic algae, and in order to ensure there is a continuous amount of nutrients being delivered to the coral, the coral regulates the number of zooxanthellae cells, as well as the amount of chlorophyll in them. Because zooxanthellae is a photosynthetic algae, zooxanthellae are sensitive to light intensity, which can ultimately alter the color and overall health of the coral. Too much light intensity can release some zooxanthellae, or the amount of chlorophyll will be decreased. In the presence of harsh light intensity, it can potentially be detrimental to the coral reef by excess oxygen production, causing it to accumulate and have toxic effects on the coral. If the light intensity is insufficient, the zooxanthellae will not be able to provide enough nutrients for the host coral, and thus the number of zooxanthellae and amount of chlorophyll will increase. When the zooxanthellae cells use light to produce large organic compounds, oxygen is also released. The brighter the color of the coral, the more photosynthesis and oxygen production. Corals can decrease or increase the chlorophyll production from the zooxanthellae cells, depending on environmental requirements. They can also expel zooxanthellae when under stress.

So, what does this mean? Essentially, the number of zooxanthellae and the amount of chlorophyll affect coral coloration. Because zooxanthellae cells range in color from a golden-yellow to brown, and when there is a large number of zooxanthellae cells present the coral color appears brown, it suggests that the light intensity affects the number of zooxanthellae and the amount of chlorophyll.

Why are some more colorful than others?

"Scuba diving Bali Nusa Lembongan, Coral Reef off Nusa Lembongan, Bali" by Wikimedia Commons under CC 2.0

Corals are home to different types, or clades of zooxanthellae, that vary in light intensity sensitivity and temperature. While zooxanthellae color can range from a golden-yellow to a brown pigment, zooxanthellae can also fluoresce deep red color under certain circumstances. For example, Phycoerythrin is a photosynthetic pigment that fluoresces a bright orange color that's found in zooxanthellae. In addition, there are about 85 more fluorescent pigment colors produced by colorful corals, typically cyan, green, yellow, and red, and they can even glow in the right lighting! These proteins absorb light of one color (wavelength) and emit (fluoresce) a different (lower energy wavelength) color. There are also other pigments produced by the coral called chromoproteins, they are non-fluorescent, or reflective, and there are about 24 chromoproteins found in corals. Of the 24 chromoproteins, the pigments can appear as purple, red, blue, to name just a few.

Coral Reef by Jan-Mallander via Pixabay [CC 2.0]

What Are the Roles of the Corals Colors?

It is thought that these fluorescent compounds may help corals survive, however, the role is not yet well understood. A few possibilities are that they may serve as a "sunscreen" protecting corals by filtering out harmful ultraviolet rays. Corals can manipulate the zooxanthellae cells in response to light. Only recently have scientists begun to comprehend the relationship between color-producing light and how it appears at different depths. Wavelengths of light become diluted the deeper down they have to travel. UV light rays are naturally filtered out by water in deeper oceanic regions. That's why shallow-water reefs are bright with color and deeper coral tend to appear more gray-like. It has also been shown that injured corals often form colorful patches. This is due to the fact that the coral is making fluorescent molecules that act as antioxidants, capturing toxic oxygen radicals that threaten to damage cells.

The information in this chapter is thanks to content contributions from Jaime Marsh and Haley Zanga

6 Chapter SYMBIOTIC RELATIONSHIPS IN CORAL REEF ECOSYSTEM

Coral reefs are home for many organisms such as sponges, fish including large nurse sharks and reef sharks to groupers, clownfish, eels, snappers, and parrotfish, jellyfish, anemones, crustaceans, other invertebrates and algae. So, how do coral reefs support such a huge weight on their shoulders? It is the symbiotic relationship that is formed when two different species interact with each other. These interactions create a balance within the ecosystem because at least one of the species is gaining from it. The other species may also gain from the relationship, be unaffected or even get harmed from the relationship. Symbiotic relationships are very common in the ocean, especially near coral reefs. There are three main types of symbiotic relationships. They are mutualism, parasitism, and commensalism. Mimicry is also frequently seen amongst coral reef organisms.

TYPES OF MUTUALISM

- There are two primary types of mutualism: obligate mutualism and facultative mutualism.
- Mutualism, or a mutualistic relationship, by definition, is when two organisms of different species work together so that each is benefiting from the relationship.
- Obligate mutualism is when neither organism can live without one another.
- An example of obligate mutualism is the relationship between ants and Acacia plants. While the plant provides shelter and food for the ants, the ants actually defend the plant from organisms such as other herbivores that may eat the plant, as well as remove any other species of plants that may limit the plant's growth.

A more specific example of obligate mutualism that is more related to this topic would be the relationship between hard coral and algae (zooxanthellae). The relationship between coral and zooxanthellae (algae), is one of the most important mutualistic relationships within the coral reef ecosystem. Zooxanthellae are microscopic, photosynthetic algae that reside inside the coral. The hard coral provides protection, as well as compounds needed for photosynthesis to occur. In return for their protection for herbivores and other organisms, zooxanthellae photosynthesize organic compounds from the sun, and then pass the nutrients, glucose, glycerol, and amino acids, which are the products of photosynthesis, to their coral hosts, essentially giving the coral reefs their beautiful colors.

"Acacia Ants" photo via Wikimedia Commons under 2.0

The corals then use those nutrients to produce proteins, fats, carbohydrates, and calcium carbonate. This is so important, in fact, approximately 90% of the nutrients produced during the photosynthesis in zooxanthellae is transferred to the coral for their use. Zooxanthellae also aid in the excretion, or removal of waste such as carbon dioxide and nitrogen. Ultimately, without algae, coral would starve to death (coral bleaching), and if algae didn't have protection, they would be more vulnerable to several herbivores and other organisms. This relationship is so important, that if this mutualistic relationship did not exist, it would be very likely coral reefs would not even exist. Therefore, making this relationship obligate mutualism, as mentioned before.

"Coral Polyp" By Emaze

The mutualistic relationship between anemones and clownfish is also another commonly known relationship. Clownfish are found in warmer waters of the Indian and Pacific oceans. Of the over 1,000 anemone species that live in the ocean, only 10 species coexists with the 26 species of tropical clownfish. Within these species, only select pairs of anemone and clownfish are compatible. Sea anemones are actually predators, with stinging polyps, that attach themselves to rocks, the ocean floor, or even coral. They patiently wait for fish to swim by close enough to get entangled in their poisonous tentacles. The toxins paralyze their prey, and the tentacles guide the prey into the anemone's mouth. However, clownfish are the exception and actually call the anemone home. Clownfish are coated with a mucus layer that essentially makes them immune to the deadly sting of the anemone. Therefore, clownfish are able to live within the anemone's tentacles, while also gaining protection from predators, and the clownfish helps feed the anemone by either letting them eat their leftovers, or by also luring fish over to the anemone, so that the anemone can catch them with their poisonous tentacles, and eat them for dinner (or maybe lunch).

"Clownfish and Sea Anemone" by Samuel Chow under Flickr

Facultative mutualism the other type of mutualism, is when species benefit from one another, but do not necessarily fully depend on one another.

An example of facultative mutualism is the relationship between certain types of our gut bacteria, or the bacteria that live in our digestive tracts, and us humans. When we eat food, bacteria use some of the nutrients from that food we are actually digesting, and in return, they help us digest our food.

Goatfish, *Mulloidichthys flavolineatus* at Kona, Hawaii is being cleaned by two cleaner Wrasses, *Laborites phthirophagus photo via* Wikimedia Commons under CC 3.0

Again, a more specific example of facultative mutualism that is more so related to the coral reef ecosystem, is the relationship between shrimp or smaller fish and large marine organisms. The shrimp or cleaning fish remove materials, such as parasites, off of the larger marine organisms, in which they get a meal from, and the larger marine organisms have potentially harmful parasites removed!

As mentioned before, earlier on in the post, smaller fish or cleaner shrimp, such as the Bluehead Wrasse or Spanish Hogfish remove parasites and other materials off larger marine organisms such as fish, sharks, and rays. In most cases, these smaller fish would typically be the larger marine organism's prey, however, in this case, these larger organisms gain the benefit of having these parasites removed, that could potentially cause harm, while the smaller fish or shrimp get a meal.

Cleaner fish and larger fish share a mutualistic relationship. This is because the cleaner fish eats harmful parasites and other small sources of food off of the large fish. This gives the cleaner fish a meal, the larger fish is helped because it no longer has these parasites on them. Often times larger fish wait in "**cleaning stations"** for the cleaner fish to come and get these things off of them. Some small shrimp can also be cleaners. The picture below shows a cleaner shrimp cleaning a large fish at a cleaning station that would normally eat the shrimp if it wasn't for this mutualism.

"Dangerous Dining" By Chris Lewis Under Vimeo

Another facultative mutualistic relationship is between the root-fouling sponge called *Tedania inis*, and red mangrove called *Rhizophora mangle.* In this relationship, the red mangrove provides the sponge with carbon that was produced by the mangrove, and the nitrogen the sponge releases gets eaten up by the mangrove to enhance growth.

Mutualism also occurs between spider crabs and algae. This relationship benefits both of these species because the greenish-brown algae live on the spiders back, which helps the spider crab blend into the shallow areas of the ocean floor where they live. In return, the algae benefit from a good place to live.

"Japanese Spider Crab" By (Ovo) Under Flickr

Another example is the relationship between the Boxer Crab and anemones. In this relationship, the Boxer Crab carries around two anemones that sting and it uses them for protection. The anemones are benefited because since the crab carries them around, it allows them to be mobile which increases their options for finding food.

"Boxing (Pom Pom) Crab" By Liquidguru Under Vimeo

One last mutualistic relationship is the relationship between a goby (*Nes Longus and Ctenogobius saepepallens*) and a snapping shrimp (Alpheus floridanus). The shrimp dig a decent sized burrow in the floor of the ocean, and the goby will then live in the entrance of that burrow. When the shrimp exits the burrow, it will stay in contact with the goby through its antennae, and depending on the species of the goby, it will either signal to the shrimp of approaching predators by darting headfirst back into the burrow or by flicking its caudal tail. Ultimately, the goby gets a free place to live and hide from potential predators, while in return the shrimp gets a look-out individual while it hunts for food!

"Goby fish with shrimp" photo via Wikimedia Commons under public domain

PARASITISM

Parasitism is not a mutualistic relationship because only one of the species is benefited. The parasite gains from the relationship while the other species involved is harmed. Ectoparasites live on the outside of the host body, whereas endoparasites live inside the host.

One example of a ectoparasitic relationship is between fish lice and small fish hosts. The fish can be killed if there are too many fish lice attached to it. The lice benefit from the fish by feeding off of their bodily fluids.

"Sea Lice On Salmon" By 7Barrym0re Under Public Domain

Isopods can also cause be involved in a parasitic relationship. Some isopods will eat the fishes tongue and then live in the fishes mouth so they can eat whatever the fish is attempting to eat.

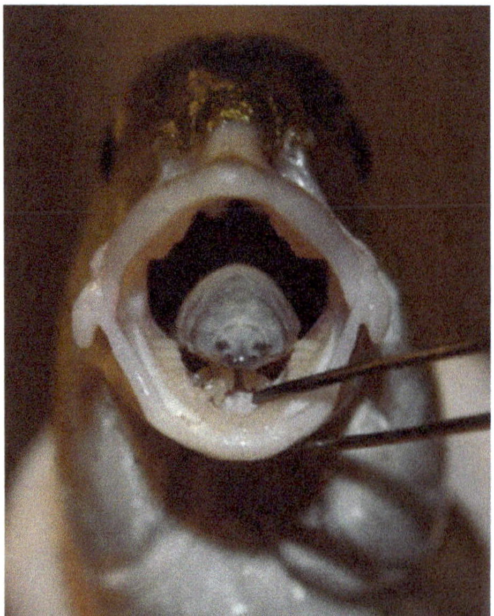

"Betty In Mouth" By University Of Salford Press Office Under Flickr

Commensalism:

Commensalism is a relationship where one species benefits from another species. The other species is neither harmed nor helped in this relationship. There are many examples of commensalism in the ocean.

One example of commensalism among marine life is jellyfish and small fish. The small fish will typically hide inside of the jellyfish's stinging tentacles if the stinging does not affect them. The tentacles provide protection for the fish from larger predators. This relationship has no effect on the jellyfish.

"Baby Fish Take Shelter In Jellyfish" By Earth Touch News Network

Another relationship is between shrimp and a featherstar. The shrimp will blend in with the featherstar and use it for protection. As you can see in the picture below, it is very difficult to find the shrimp hiding in there.

"Shrimp in Featherstar" **by** prilfish **under Flickr**

Mimicry:

One type of Mimicry is when one organism that is harmless evolves to look similar to another organism that is poisonous. This stops predators from eating them because they think they are the poisonous species. They can also use mimicry to appear larger than they really are.

The four-eye butterflyfish uses a large eyespot in order to appear larger to predators.

"Chaetodon Capistratus1" **By Chris Huss Under** Public Domain

Another example of mimicry is between the Sabre-tooth Blenny and Cleaner Wrasses. The Cleaner Wrasse have a mutualistic relationship with larger fish so they don't get eaten, and the Sabre-tooth Blenny takes advantage of this relationship by evolving to look very similar to the Cleaner Wrasse. Instead of cleaning the larger fish, the Sabre-tooth Blenny will take a bite out the the large fish's flesh and swim away. This is an example of aggressive mimicry.

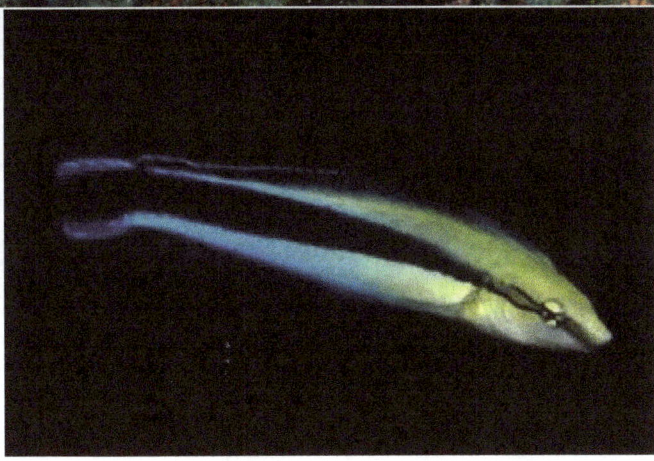

"Bluestreak Cleaner Wrasse" By Nemo's Great Uncle Under Flickr "Sabre-Tooth Blenny" By Fish Index

On the top, there is a Bluestreak cleaner wrasse and on the bottom is a sabre-tooth blenny. You can see how similar they look and how fish could mistake them.

WHAT'S THE BIG DEAL?

"General characteristics of a large marine ecosystem (Gulf of Alaska)" photo via Wikimedia Commons under Public Domain

Mutualistic relationships, whether obligate or facultative mutualism, are an integral part of sustaining a coral reef ecosystem, and without them, the coral reefs would simply not exist. These mutualistic relationships define a largely intricate number of connections and relationships which deeply rely on one other, and where one could start to deteriorate, another could as well. It is more important now than ever that we sustain healthy coral reefs to support these intricate relationships, communities, and ecosystems.

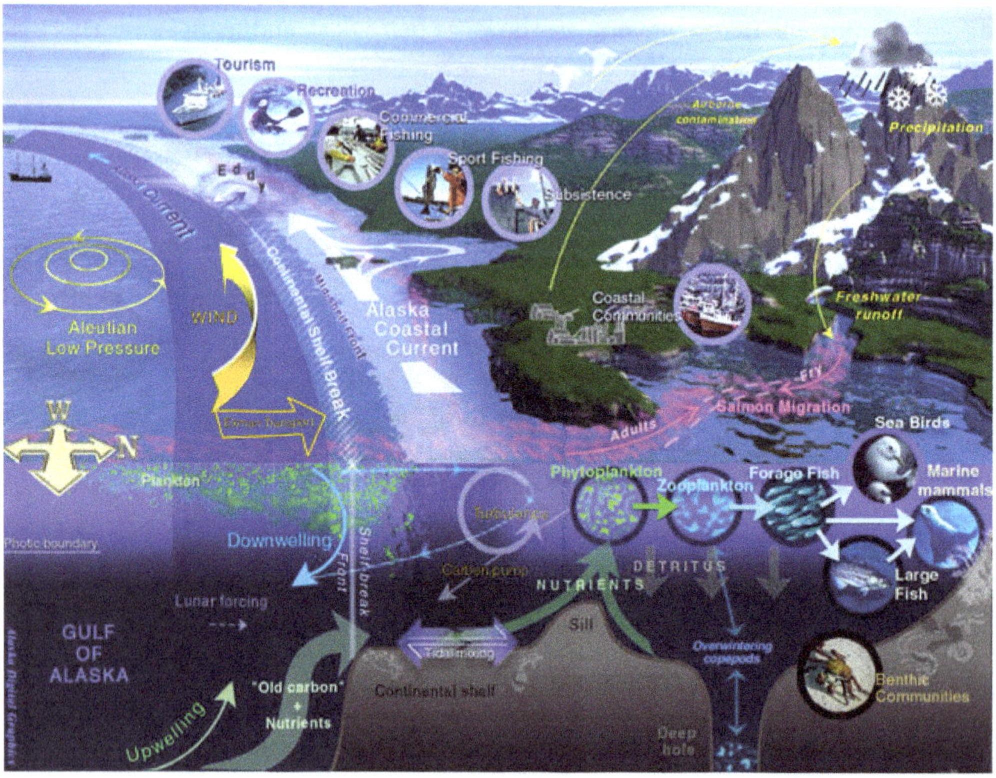

The information in this chapter is thanks to content contributions from Jaime Marsh, Christian Paparazzo, and Alana Olendorf

7 Chapter

IMPORTANCE OF CORAL REEFS

WHY ARE REEFS IMPORTANT?

Coral reefs are very complex ecosystems that provide valuable habitat for fish and other animals with their beautiful and unique structures. These structures provide shelter for many organisms such as fish, marine worms, clams and many other animals and plants that all play a vital role in the coral reef ecosystem. Coral reefs are important for a variety of reasons which we will discuss below.

Coral Outcrop Flynn Reef by Toby Hudson via Wikimedia [CC by 2.0]

Biodiversity

Coral reefs are often thought of as a busy city; the buildings being made of coral and the thousands of organisms inhabiting this city acting like the humans interacting with each other and performing daily jobs. Coral reefs provide protection and shelter to nearly one-quarter of all known marine species and have evolved into one of the largest and most complex ecosystems known to humans. Coral reefs are home to over 4,000 species of fish, 700 species of coral and thousands of other plants and animals. This diversity of species provides a large gene pool giving communities more resilience during extreme environmental conditions and climate change. This is important to the overall health of an ecological community. With greater species diversity, the impact of losing any one species to extinction will be less. The enormous diversity of coral reef organisms also provides potential for new medicines or other products that may be developed from biochemicals that these organisms produce. Most coral reef organisms have not been studied for their potential benefits to medicine and industry.

Coastal Protection

Coral reefs act as a natural barrier protecting coastal beaches, cities, and communities from the waves of the ocean. Nearly 200 million people depend on coral reefs to protect them from storm surges and waves. Without coral reefs, many buildings would become vulnerable to storm damage.

Reefs Protect Costallines by NOAA ***via*** Flicker **[CC by 2.0]**

Food

Coral reefs are an important food source for the people who live near the reefs and are crucial for worlds fisheries providing them with a significant source of protein. In developing countries, the reef is said to contribute to one-quarter of the total fish catch providing food resources for tens of millions of people.

"Commercial Fisherman" **by** NOAA **[CC by 2.0]**

Medicine

Many of the compounds now being used in human medicines are found on the coral reef with the potential of more to be discovered. A number of organisms found on reefs produce chemical compounds that have been isolated for human applications. Scientists have developed treatments for a variety of illnesses such as cardiovascular diseases, skin cancer, ulcers, and leukemia. Other compounds can help with reducing inflammation, kill viruses and relax muscles. Not only do the organisms inhabiting the coral reef provide medical treatments but the coral's unique skeletal structure has been used for bone-grafting material.

Tourism

Coral reef ecosystems are among the most biologically diverse and economically valuable ecosystems on Earth, as they not only support local but global economies. Through tourism (*i.e.* snorkeling, scuba diving, swimming) and fisheries, coral reefs generate billions of dollars, as well as jobs in more than 100 countries around the world. The annual value of the ecosystem services provided by coral reefs to millions of people is estimated to be over $375 billion. The coral reefs can indirectly bring economic value to these countries by letting visitors enjoy beaches, eat local seafood, paddleboard and sail. All of this is possible due to coral reefs acting as a buffer against waves, storms, and floods. The downside is when tourism harms the coral reefs it not only affects the organisms and the coral reefs, but local people and their economy who rely on the income from tourism.

Snorkel by Angelique800326 **via Wikimedia [CC by 2.0]**

A good example can be found in Bonaire, a small Caribbean island. Bonaire earns about 23$ million (USD) annually from coral reef activities, yet managing its marine park costs less than $1 million per year. A study conducted in 2002 estimated the value of coral reefs at 10$ billion, with direct economic benefits of 360$ million per year. For residents of coral reef areas who depend on income from tourism, reef destruction creates a significant loss of employment in the tourism, marine recreation, and sport fishing industries. This large amount of money of revenue generated is being threatened by the degradation of coral reefs. As you can see there is a positive feedback loop occurring because of this situation. Many components of tourism, including recreational activities, are the cause of damage to the reefs but ironically it has been shown that ecotourism is as well.

8 Chapter

CORAL REEFS ARE COMPLEX ECOSYSTEMS

CORAL REEFS AS ECOSYSTEMS

An ecosystem includes all of the living organisms (biotic) in a given area, interacting with each other, and also with their non-living components and environmental factors (abiotic). In an ecosystem, each species has its own niche or role to play. All of the aspects of the coral reef act together in a unit called an ecosystem. Coral reefs are a very high functioning ecosystem and are home to thousands of species of marine life. Algae, fish, echinoderms and many other species depend on the reef for their habitat and food too. The number of species is directly proportional to the mass of the coral reef. Mesophotic coral ecosystems (MCEs), are a common type of ecosystem that are home to light-dependent life, such as corals and their zooxanthellae. MCEs are typically at depths of 40-30 m and can extend over 150 m in tropical and subtropical regions.

ROLES OF MARINE ORGANISMS

The biotic portion of the marine ecosystem includes three main groups called producers, consumers, and decomposers. The organisms in these groups all play a key role in contributing to a functioning ecosystem.

- **Producers–** These are autotrophic organisms which make their food through photosynthesis. Green plants, algae, and chemo-synthetic bacteria are all examples of producers in marine habitats.
- Consumers– These organisms obtain food from other organisms or organic matter and are animals, zooplankton, and heterotrophs. Consumers are broken down into primary, secondary, tertiary, and quaternary categories. Primary consumers feed on producers and are herbivores. These organisms include sea turtles, zooplankton, and sea urchins. The secondary level feed on the primary producers and are organisms like rays and fish. The last levels are tertiary and quaternary, which feed on the secondary consumers and are the large fish, sharks, and sea lions.

- Detritivores and Decomposers– These organisms feed on dead organic matter. Crustaceans like crabs and lobsters which shred and consume dead animals are examples of detritivores. Bacteria and fungi are decomposers that gain energy by breaking down dead organic matter into nutrients such as nitrates, nitrites, phosphates, and carbon dioxide. These are released back into the surrounding environment along with heat energy.

The health, abundance, and diversity of the organisms that make up a coral reef are directly linked to other surrounding terrestrial and marine environments. Mangroves and seagrass beds are two of the most important associated habitats of the greater coral reef ecosystem.

Mangroves

Mangroves are salt-tolerant trees that grow along tropical and subtropical coasts. They provide protection by stabilizing the coastline with their complex root system. They protect uplands from erosion, wind, waves, and floods. They also act as carbon storage systems, and help produce nutrients and filter out pollutants. Their complex root system serves as breeding and feeding grounds for marine organisms such as fish, invertebrates, and others. They are especially important as **nursery grounds** for juvenile and larval reef organisms, many of the animals raised in mangroves migrate to coral reefs.

There are around 80 different species of mangrove trees that live in many places around the world. In Florida, the Bahamas and the Caribbean, there are three main types of mangrove. Red mangroves grow on the ocean's edge, black mangroves occur a bit upland from the reds, and white mangroves are found at the highest elevations furthest upland.

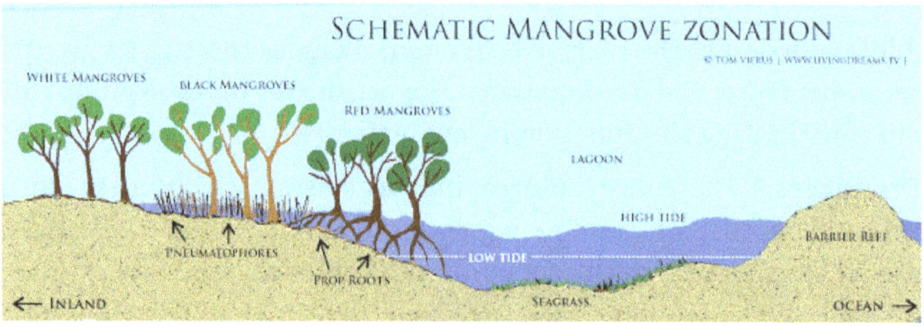

The schematic mangrove zonation done by Tom Vierus **on www.livingdreams.tv.**

In Mumbai, the shores once used to be covered in mangroves but those forests were severely degraded from housing developments encroaching on the edge of the land and mismanagement of waste disposal. Mumbai alone has lost about 40% of its mangrove forests in just a few decades from being destroyed for houses to be built or also from the wastes left behind which can clog the roots of the trees preventing them from getting oxygen and filtering the freshwater they need.

Mission Mangroves, a project run by NGO United Way Mumbai, is making changes towards protecting the remaining mangroves and planting new ones. Every month volunteers partake in a clean up of the wetlands and twice a week there are planting sessions to restore the forests. At the time of this writing, they have already planted 55,000 new trees. United Way Mumbai's director of community impact, Ajay Govale, stated that "the mission is to green 20 acres of wetland by planting mangroves, and to educate thousands of Mumbai people about the importance of mangroves."

India is at the top for countries that are the worst at mismanaging plastic wastes and Mumbai is one of the worst cities for air pollution. The destruction of mangroves has contributed to this problem as they help filter pollutants out of the air and can absorb four times more carbon than tropical forests. The loss of mangrove forests also impacts other parts of the ecological community. Since they serve as shelter and breeding grounds for marine wildlife, this affects local fisheries and communities reliant on these marine animals. The loss of the mangrove forests also leaves the coastlines more vulnerable to be affected by rising sea levels from global climate change.

Seagrasses

Seagrass: Seagrasses are flowering plants that often form meadows between mangrove habitats and coral reefs. They provide nutrients to organisms such as sea turtles, sea urchins and thousands of other species. Seagrasses also provide protection and shelter for crustaceans like crabs and lobsters, as well as fish such as snappers. Similar to mangroves, they also function as nursery grounds for juveniles, perform water filtration, release oxygen necessary for most marine life, and prevent seabed erosion.

Seagrass Meadows **by** U.S. Department of Agriculture **via Flicker**

9 Chapter

LARVAL DISPERSAL AND SETTLEMENT

Many marine organisms have at least a single larval stage. The reproductive adults will release many offspring into a water column. These free-swimming larvae will then become part of the zooplankton, being carried mostly by currents. This life stage is very important to some species, who as adults are benthic. Once the larvae settle, they are there for life. There are three methods of larval dispersion and development:

Direct development or crawl-away larvae have a low dispersal potential. The young usually hatch from the egg looking very similar to the adults of the species.

Lecithotropic larvae have more of a dispersal potential than crawl-away larvae. One thing that characterizes this type is that they are provided with a yolk sac, or some other form of nutrition. This finite source of nutrition only allows for a certain amount of time to disperse and settle before this nutrition source runs out.

Planktotrophic larvae have the greatest dispersal potential. they can survive as pelagic larvae longer than the other types of larvae. This is because they are able to feed on smaller zooplankton and phytoplankton. Most sessile invertebrates have planktotrophic larvae.

There are many settlement cues, all of which vary from species to species.

Chemical cues are common and heavily studied. These cues are biological compounds created by another individual that larvae pick up on and use as a way to tell if a location is safe to settle in. These cues can be from adults of the same species, which ensures it is a safe place to settle, or from predators, which ensures its not a safe place to live. Raymond C. Highsmith studied the induced settlement in sand dollars, *Dendraster excentricus*. He found that the larvae of *D. excentricus* showed a preference for adult-associated sand or sand with adults present. He also found that the type of substrate didn't matter, as long as adults were present, thus proving a chemical cue. The preference for this settlement cue is most likely due to the absence of micro predators that feed on metamorphizing larvae, such as *Leptochelia dubai*, from areas where adults are present. The reworking of the substrate by the adult sand dollars makes it impossible for these micro-predators to live there, leaving it a safe place to settle.

Sand Dollar by mosaikweb Via Pixabay

Physical settlement cues are also important for some species. Some larvae may only settle in areas with certain substrate types. Fish larvae are commonly seen to use physical cues as they prefer certain habitat types. Whalan et al. looked at this largely ignored cue. They studied five types of sessile marine invertebrates found on coral reefs, two species of scleractinian coral and three species of sponge. They found that both of the coral and one sponge species had significantly higher settlement on tiles with microtopography that included divots that closely matched the larval width. This proves that physical cues also play a role in the settlement.

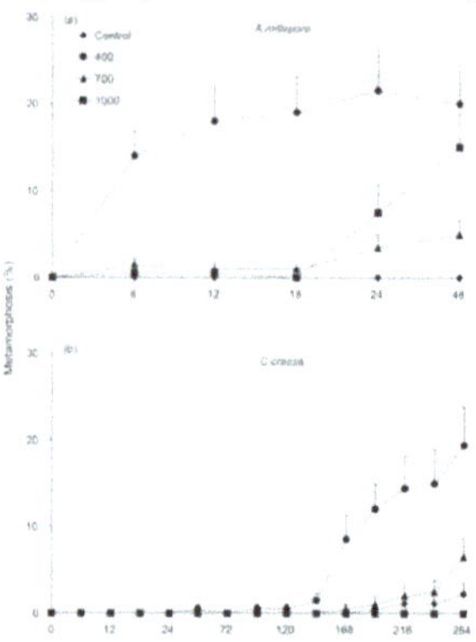

Figure 3 from Whalan et al. [CC BY 4.0] shows the time in hours taken for a percent of coral larvae to settle and metamorphosize. Each line is a different sized divot, and the control was a smooth tile.

Previously there have been observations about where coral reef fish larvae are orientated when they swim offshore. They require orientation cues. A study done by Jack O'Connor and Rachel Muheim examined the effects of magnetic fields on the orientation of coral reef fish. During the observations, coral-reef fish larvae revealed remarkably consistent orientation behavior while swimming offshore, requiring large-scale orientation cues. However, the mechanisms causing this behavior are still being investigated. One potential large-scale cue for orientation is the Earth's geomagnetic field. Here, they examined the effect of magnetic field manipulations on the orientation behavior of coral-reef fish during the pelagic larval phase. In the absence of visual cues, individual larvae responded to a 90-degree shift of the horizontal component of the magnetic field within a Helmholtz coil with a comparable shift in orientation. This shows that they use a magnetic compass for orientation. Their findings suggest that geomagnetic field information guides the swimming behavior of larval fish in the pre-settlement phase. The ability to use large-scale sensory cues allows location-independent orientation of swimming, a behavior that influences dispersal and connectivity of fish populations, which has important ecological implications for anthropogenic development of marine areas.

10 Chapter | SQUIRREL FISH

The fish species Holocentrus adscensionis, also known commonly as the squirrelfish or soldierfish, is found in the family Holocentridae. There are around 70 different species of the tropical reef squirrelfish. They have large eyes to help them see at night, are colorful (typically red and gold) with spiny elongated fins. While they can grow up to two feet in total length, they commonly do not surpass 10 inches. Squirrelfish are found distributed throughout the warm tropical waters of the Atlantic Ocean around the Caribbean, Bahamas, Florida, Bermuda, Turks and Caicos, and the Gulf of Mexico, where they typically remain at depths four feet to 40 feet where the waters are still warm but can go up to around 250 meters deep.

Holocentrus rufus by Nick Hobgood via Wikimedia

Squirrelfish have all five fins, a see-through pectoral fin, ventral, anal, and elongated dorsal and caudal tail fins. They also have fin spines along their spine with horizontal striped white lines along their back below them. In some species of squirrelfish, they have spines on their gill covers that are venomous that they use for self-protection. When the squirrelfish are juveniles they have more iridophores, cells that reflect light, which give them a silvery shimmer. When they transition into adults is when the red pigments of the chromatophores are more prominent distinguishing the young from the adults. The red orangey color helps them to blend in with the corals they sleep in during the day. The IUNC red list has classified squirrelfish globally as a species of least concern when they were assessed in January of 2013 but, global warming may have changed this trend as the corals are being bleached from being stressed as a result from warming ocean waters.

Squirrel-fish by Skinned Mink via Flickr

Squirrelfish are nocturnal carnivorous fish that hide in crevices of the coral reefs during the day to avoid predation. At night they swim through the reefs through seagrass beds hunting meroplankton, larvae and small crustaceans with the occasional small fish. When these fish are young, they tend to group together with one another which helps with protection and hunting, while adults prefer to establish their own territory and be alone. They are also able to communicate intra-specifically by producing sounds with their swim bladders. They make these sounds through vibrations to warn off predators or define their territory.

The information in this chapter is thanks to content contributions from Emma Verville

11 Chapter

JACKKNIFE FISH

The Jack-knifefish *(Equetus lanceolatus)*, is a black and white/silvery fish which belongs to the *Sciaenidae* family, the drums category. It is a Carribean reef fish and can be found in the Carribean, Bahamas, Florida, and the Gulf of Mexico.

Photo by Barry Peters **is licensed under CC BY 2.0. This is a juvenile Jackknife fish.**

Morphology

Morphology deals with the size, form, shape, and structure of organisms. The Jack-knifefish is a fairly small fish ranging from about 5-9 inches in size (12-23 cm). It has an elongated dorsal fin with a black band that runs from the tip of the dorsal fin to the end of the tail, which helps identify this exotic looking fish. They are an odd-shaped fish and have considerably long dorsal and caudal fins. The combination of those two fins resembles a jack-knife, giving it it's a common name, the Jack-knifefish. This odd trait is more present in a juvenile jackknife fish than older jackknife fishes.

General description and Behavior

The Jackknife fish is a shy and graceful saltwater fish, so they require a peaceful environment with a sandy bottom and plenty of places/rocks to hide in. They feed on inhabitants of reefs like ornamental shrimps, polychaete worms, and even other small reef fishes; because of this, I think it's safe to say that they are carnivorous benthic feeders.

The odd shape of the Jackknife fish actually serves as a kind of protection from predators by confusing them into thinking that it is two different fishes, rather than one. The coloration of the jackknife fish may also serve to hide the eyes so that predators cannot tell where the fish is looking, and even confuse the predators into thinking maybe it's not even a fish.

The information in this chapter is thanks to content contributions from Malisa Rai

12 Chapter
QUEEN TRIGGERFISH

Queen Trigger Fish on the Fathom **by** Johnmartindavies**under** CC **by 2.0**

Specific Name: Balistes Vetula

Genus: Balistes

Species: Vetula

Family: Balistidae

This beautiful Queen Triggerfish can be described to be a large oval shape that is laterally compressed. The eyes are located towards the top of its head. Their eyes have the ability to move independently of each other. They are about 8-16 inches long. The name 'triggerfish" comes from the two spines on the anterior dorsal fins that lock the fish into crevices at night. How does she do this? Good question! The Queen Triggerfish uses the first fin to lock itself into the crevice and the second fin to unlock itself. This is a defensive mechanism to keep themselves from getting eaten by their predators. This isn't the only way they protect themselves; these amazing fish can also produce a throbbing sound by using a specialized membrane located beneath their pectoral fin. This throbbing sound is a warning to their predators to stay away.

Coloration:

The back of theQueen Triggerfish is typically green or blue while its abdomen and the lower head are orange and yellow. One of the prettiest features of this fish is the bright blue bands that extend from the snout to the front and below the pectoral fins, the bands around its mouth, and the bands around the caudal peduncle and median fins. Both female and male have similar morphology. The difference in coloration differs in juveniles because the younger triggerfish are paler in color and have smaller fins.

Queen triggerfish **by** adam **under** CC **by 2.0**

Where can I find them?

The Queen Triggerfish can be found in depths of approximately 7-902 feet, typically over rocky bottoms and around reefs. They also form schools. Geographically they are located in the Caribbean, Bahamas, Florida, Bermuda and the Gulf of Mexico.

What's for dinner?

The Queen Triggerfish are primarily carnivores but are also sometimes herbivorous as they occasionally eat algae. The majority of their diet consists of sea urchins, bivalves, crabs, starfish, sea cucumbers, shrimp and polychaetes.

The information in this chapter is thanks to content contributions from Jennifer Rosado

13 Chapter | BLUE CHROMIS

Chromis cyanea *by* Wikimedia Commons *under* CC 4.0

GEOGRAPHIC LOCATION AND HABITAT

Chromis cyanea or Blue Chromis are located in the Western parts of the Atlantic Ocean including the Gulf of Mexico and the Caribbean Sea and are also found on the coast of Bermuda. They are most commonly seen swimming in reef habitats but can also be living in lagoons where food and shelter are prevalent. They are most commonly found swimming at depths of 10 meters to 20 meters. These fish rely greatly on the health of coral as they provide the Blue Chromis with a place to live, breed, feed and also protection from potential predators.

Description

This oval-shaped reef fish is about 13 to 15cm in length once fully grown in their adult life stage. These fish get their name from their bright blue body color. They also have a black striped dorsal fin, anal fin, and caudal fin. One fun fact about Blue Chromis is that they have dark eyes which are a great way for people to distinguish this species among other similar ones.

Threats

These fish face several threats in their everyday life, some of which are not often touched upon by many people and they slide under the radar. One of the most serious threats is the Chromis trade. Large numbers of Blue Chromis are collected for use in aquaria. Damselfish or Blue Chromis make up a very large portion of the life we see in aquariums around the globe, almost half in most portions of the world. Another, possibly less serious threat is predation from lionfish which are growing in number every day. These invasive fish consume small damselfish and other reef fishes in the Caribbean and Atlantic oceans. Finally, another large and growing problem is the loss of coral which occurs in all oceans. The coral reefs are what provide the structure for the entire ecosystem which the Chromis thrive in and spend almost all of their lives.

The information in this chapter is thanks to content contributions from Devon Audibert

13 Chapter

HONEYCOMB} COWFISH

Honeycomb Cowfish (Acanthostracion polygonius) belong to the Ostaciidae family, typically known as boxfishes. The structure of fish in the family include many bones and are shaped like a square. They have small mouths and broom-like tails that help with swimming because the boxy shape weighs them down. The Honeycomb Cowfish, however, is a rare species and have heavy hexagonal scales that cover the whole body. One form of defense are the tiny horns that protrude over the eyes. They typically grow between the range of 7 to 15 inches and are commonly found in water depth of 20-60 ft, meaning they inhabitant the neritic zone.

HoneyComb Cowfish CC BY SA 2.0

Location:

The habitat of Honeycomb Cowfish is the western Atlantic Ocean. They are found as far up along the coast of New Jersey and as far south as southern Brazil, but are most typically found in the Caribbean and Gulf of Mexico. These diurnal animals prefer to reside in clear water with an average temperature of 22-270 Cin coral reef habitats and vicinity.

Anatomy:

A closer look at the biology of the fish shows the hexagonal scales are attached firmly to each other except around the head and tail of the fish. This allows for respiration and movement for the gills, fins, and eyes, and caudal peduncle. The color of the fish is also a unique characteristic; they are seen in blues and greens and yellows and browns. The dorsal edge has darker and/or irregular hexagons due to the opening of the scales. Juvenile fish are more brightly colored and have shorter dorsal fins. The honeycomb scales are used for protection. Honeycomb fish feed on marine invertebrates, mainly shrimp, tunicates, and sponges. The structure of the small mouth is used to suck in small food particles. Their food sources are mainly sessile which are easier to find, as honeycomb cowfish are not active hunters. They have a solitary nature and are never seen in groups of more than 3 individuals. Typically, when grouping, the small school will be comprised of one male and two females. Not much is known about their reproductive nature, but many scientists conclude this grouping may be significant. It is known that they are open water mating fish which means they swim to the surface quickly to release gametes and then swim back down immediately.

Cowfish Honeycomb By Amanderson2 **CC BY SA**

Like many fish, they undergo different life stages. For Honeycomb Cowfish they have two stages of juvenile and adult. Juveniles are rounder and brighter in color. They can also change color to protect themselves from predators. They do this by sending a signal via nerve impulse to Chromatophores, which are pigment cells in the scales. To learn more about this process watch this video.

Predators of Honeycomb Cowfish are larger fish, but usually cowfish are undesirable as food due to their external armor. As Honeycomb Cowfish grow into adulthood their colors fade and they become more triangular and rigid resulting in awkward swimming. One way they protect themselves is by use of camouflage by blending into the colorful surrounding of coral reefs. When stressed, adults can brighten their colors to be more effectively hidden. When using camouflage, they remain stationary for long periods of time.

Overall, Honeycomb Fish are very unusual fish and use honeycomb-shaped scales as protective armor. They can also use their bright colors to blend into coral reefs to hide from predators and the boxy shape makes them undesirable to natural predators. Unfortunately, these fish are used to a great extent commercially as pets which can damper population sizes. By choosing their habitat close to coral reefs, they have easy access to food and shelter. These fish are really interesting. If you want to learn more about boxfishes, watch this video.

13 Chapter SPOTTED DRUM

The Spotted Drum or *Equetus punctatus* is a species of marine fish belonging to the Sciaenidae or Drumfish family. This family is named for the low resonance noise they emit that sounds like the beating of a drum. This noise is made when the fish beat their abdominal muscle against their swim bladder. It is hypothesized that this is used as a tactic to attract a mate for spawning purposes.

Spotted Drum Equetus punctatus **by** Brian Gratwicke **under** CC BY 2.0

The Spotted Drum can grow to be anywhere between six and ten inches and possesses a distinctive black and white pattern consisting of both dots and stripes and has an unusually long dorsal fin as can be seen on the adult pictured above. As juveniles, they are extremely elegant and beautiful, their patterning consists only of stripes, and they possess an extremely long dorsal fin that becomes shorter as they age.

There is not much in the literature about the purpose of their patterning specifically, but we can draw from information regarding other fish with similar patterns. For example, Angel Fish which are also solitary, possess a similar solid dark banded pattern. This pigment switch from light to dark is a mechanism to confuse predators during an attack.

Juvenile Spotted Drum **by** Laszlo Ilyes **under** CC BY 2.0

This fish can also be classified as an odd-shaped swimmer. Odd-Shaped Swimmers are often characterized by a slow and awkward swimming technique. In character with this, Spotted Drums swim around in small repetitive circles. They are solitary fish and are often described as secretive- they are very rarely seen in groups. It does not seem as though they are aggressive or territorial fish because they are very easily approached by divers.

The Spotted Drums spend their days hiding in caves and crevices in reefs in the Bahamas, Bermuda, the Caribbean, and Florida. You will on occasion spot one swimming around the base of the reef during the day. Normally they only come out at night in order to feed on things like small crustaceans, worms, and shrimp. As for predators of the Spotted Drum, according to the literature they are threatened by very little.

13 Chapter | PEACOCK FLOUNDER

Bothus lunatus, also known as the peacock flounder or plate fish is in the Lefteye Flounder (Bothidae) family. It is one of the most common flounders in coral reefs! They are usually 6 to 8 inches and thrive at a depth of 2 to 100 meters. They are found in the Bahamas, Caribbean, Bermuda, Gulf of Mexico, and Florida. *Bothus lunatus* is the Atlantic species of peacock flounder and *Bothus mancus* is the indo-pacific peacock flounder. This chapter is going to focus on the Atlantic species of peacock flounder, *Bothus lunatus* although there are many similarities between the two.

Peacock Flounder- Bothus lunatus **by** prilfish **via Flickr**

The peacock flounder is usually found in sandy areas in mangroves, seagrass meadows, and coral reefs. As a benthic organism as it spends most of its life on the ocean floor or slowly swimming slightly above.

They prey on small fish, which make up about 85.7% of its diet. However, it occasionally preys on octopi and small crustaceans such as marine shrimps and mollusks. They use their incredible camouflage to blend in with their surroundings to catch their meals. To catch their meals, they lie on the seabed partially submerged in the sediment and ambush their prey. When they aren't on the seafloor, they remain close to the sediment and swim using short gliding motions with occasional bursts of fast motion when trying to avoid predators. As a diurnally active fish, they are most active during the day and typically rest at night.

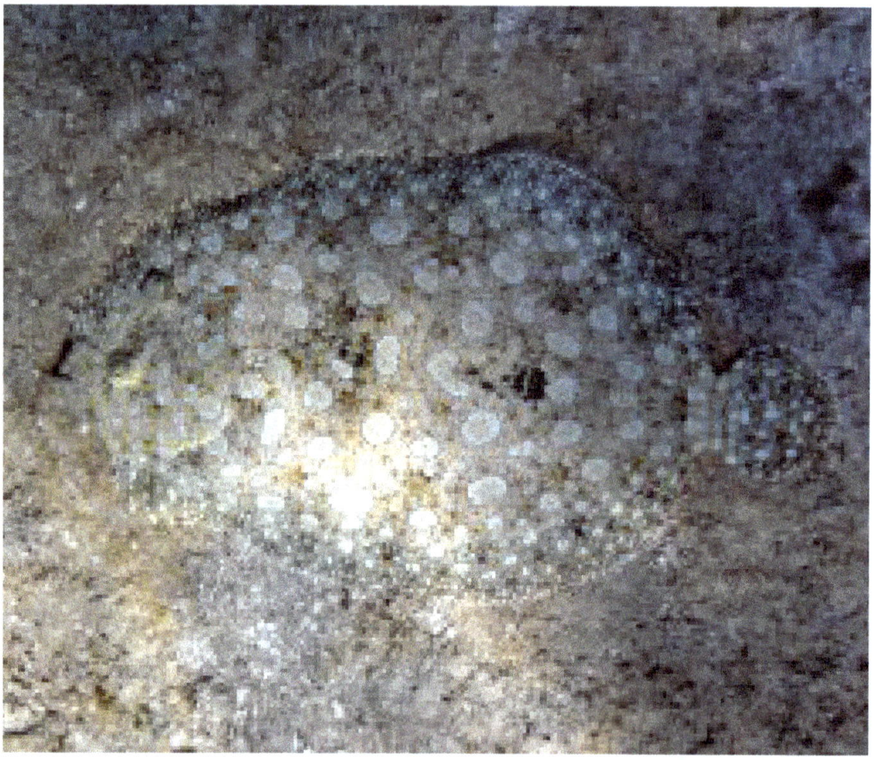

Bothus lunatus (peacock flounder) (San Salvador Island, Bahamas) by James St. John via Wikimedia Commons

As juveniles, their eyes are on opposite sides of their head, and as they mature their right eye migrates towards the left. Having both eyes on the same side of the head helps when they lay down in the sand. They can see prey better as they have two eyes facing upwards and none facing the sand. They are great at camouflage and use their eyes to see their background and adjust their color to it. A problem arises if they have damaged sight when they try to camouflage because they can't see the surroundings correctly. Their eyes are also spaced very far apart with males having an even wider gap between the left and right eyes.

When they are not camouflaged, their natural coloring is brown/grey/tan with bright blue, circular spots on their entire body, including their fins.

A colorful peacock flounder showing off its vibrant colors by Hectonichus **Via Wikimedia Commons**

Fun fact: it takes the peacock flounder somewhere between 2 to 8 seconds to *completely* blend into their background. Different ages of peacock flounder face different kinds of predators. Juvenile peacock flounder face predation from shrimp, crab, and other fish. Adult flounder are prey for a variety of animals; striped bass, cod, bluefish, groupers, moray eels, stingrays, sharks, and more. The peacock flounder can live up to 10 years and breeds year-round. They always mate just before sunset and the mating lasts for a quick 15 seconds, on average.

The peacock flounder has been evaluated by the IUCN and, as of 2012, is of least concern when it comes to extinction likelihood. However, the peacock flounder is eaten by humans and overfishing and by-catch could potentially push it into a near threatened or vulnerable status in the future. It is said that by 2048 there will be 'fishless oceans'. Now, 'fishless' doesn't mean no fish at all, at least immediately, but instead refers to a 'collapse' of the species... A species collapse means that 90% of the species are *gone*. With a 90% reduction in organisms, eventually, the number will end up climbing to 100%. This is mostly due to the issue of overfishing and the impact of climate change on ocean ecosystems. As of 2018, fishing industries are taking in 2 to 3 times as much fish than the oceans are able to support. This has caused about 85% of the WORLDS fish populations to be driven to near extinction or put on a fast track to extinction.

17 Chapter

CARIBBEAN REEF SHARK

Caribbean Reef Shark (*Carcharhinus perezi)* are in the family **Carcharhinidae.** They are most abundant in the South-Western Atlantic ocean, although they can be found from the coast of Georgia to the southern coast of Brazil.

Carribean Reef Shark **by** Leineabstiegsschleuse **via Wikimedia Commons**

SIZE, SHAPE, AND OVERALL BODY FORM OF CARIBBEAN REEF SHARK:

Size: 6ft – 10ft Adult Size Shark

Color: Silvery Grey and Greyish brown with a white underside.

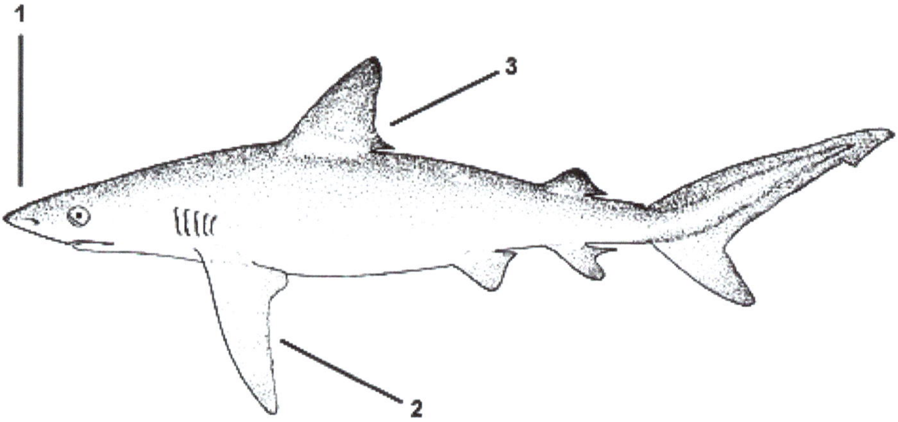

Caribbean reef shark (*Carcharhinus perezi*). Illustration courtesy FAO, Species Identification and Biodata

Shape :

- Snout is rounded defined by the short features
- Pectoral fins are long in length and skinny in appearance
- The dorsal fins are used to navigate and are short as well

Conservation status:

The Conservation of this species is monitored by International Union for Conservation of Nature. This organization rates the level of which a species is threatened in the wild.

Habitat and Behavior :

These sharks prefer to swim about the reef near the bottom hunting small boney fish in isolation. The Caribbean Reef Shark has evolved to hunt at the bottom depths having an extrasensory gland that allows these sharks to hear extra low frequency sounds making hunting for panicking fish easier. In Brazil, these sharks have been documented to hide in small caves to hunt prey and to rest. This is one of the only species of sharks that lie motionless to sleep versus counterpart species such as great white sharks that swim while sleeping.

Human interactions:

Caribbean Reef Sharks are not deadly to humans typically but will be if they feel threatened or are provoked. If you are more curious about the number of attacks each type of shark species has on humans and if they were deadly check out this link: Shark attack records

Tropical Role of Caribbean Reef Sharks :

Like many sharks, the Caribbean Reef Shark is an apex predator. They fear nothing and eat a healthy diet of cephalopods and small boney fish. They also have a mutualistic relationship with smaller fish that almost piggyback on the shark. They swim close to the bottom jaw and assist the sharks in sometimes finding prey and getting the scraps of the kill. In addition, surgeonfish, goby, and other cleaner fish will pick off algae or any type of parasites growing on the sharks when the sharks rest near sites containing these species.

Want to see these sharks in the wild check out this video on youtube of Caribbean Reef Sharks in the wild: Nature Habitat of Reef Sharks

18 Chapter | BLACK DURGON

Black Durgons, along with other Triggerfish, are in the *Balistidae* family. *Melichthys niger* is commonly referred to as the Black triggerfish. They are most commonly known for their incredible opalescent brilliance in direct light. Despite this, in the absence of light, the organism may seem dark and black, much to its evolutionary advantage. Thus it has been labeled as the Black Durgon.

Melichthys Niger. Photo by John Martin Davies, July 25, 2007 (CC-BY-SA-3.0)

Melichthys niger from adolescence to adulthood may measure 6-14 in. and could be found at depths 10-200 ft. (Reefguide). Their distribution has been described as circumtropical. They express a small to medium-oval morphology with a pronounced dorsal and anal fin. There is typically seen a distinctive blue-silver lining along the connective length of the fin to the main body of the individual. Pectoral fins are small relative to the size of the species and express the shape of a seashell.

The habitat of Black Durgon is restricted to the Neritic zone, and they are known to be mostly mobile as well as solitary in their behavior, perhaps due to their aggressiveness. Maximum lifespans have been estimated up to 11 years (Environmental Biology of Fishes). Black Triggerfish favor temperatures between 72-78o Fahrenheit and a pH of 8.1-8.4 (slightly basic).

Black Durgon over a patch reef. Photo by James St. John, June 23, 2010 (CC-BY-2.0)

The home range of these individuals is about 20o north and south of the equator in tropical regions around the globe (Black Triggerfish). They are most common to Florida, the Bahamas and the Caribbean. The Black Durgon is known to be an omnivore. This species eats essentially anything that they can procure including but not limited to: other small fish, small cephalopods, zooplankton, algae, and plants. Intriguing aspects of *Melichthys niger* are reflected in their ability to undergo biochemical processes to change color and iridescence. Examples of individuals turning stark white from their typical black-opal shade have been recorded and was likely an event to match the fish to its surroundings. Conspicuous blue streaks and markings can also be seen around the heads of these animals (Rolling Harbour). There are no obvious distinctions between the males and females of the species. When triggerfish get together for mating periods it is thought to be dictated by the moon and other tidal factors. Males are territorial in nature at this time as they build nests adequate enough to house thousands of eggs produced by the females. During gestation, male and female triggerfish are known to share care responsibilities of their offspring, fanning the eggs to ensure proper oxygen supply (National Geographic). The population status of Black Durgon is not currently listed as critical. They are renowned for their frequency, beauty, and contribution as one of the lesser aggressive forms of triggerfish, making them pleasurable to study and observe.

19 Chapter

BULL SHARKS

Brian Gratwicke – sandbar shark,
Carcharhinus plumbeus,CC BY 2.0

Bull sharks are commonly found in warm, coastal areas in freshwater or saltwater. They are well known for their aggressive behavior and can be a big risk for humans since they are known for living in high density near the shoreline. Many experts consider Bull sharks to be the most dangerous shark in the world. They are among one of 3 species most likely to attack humans.

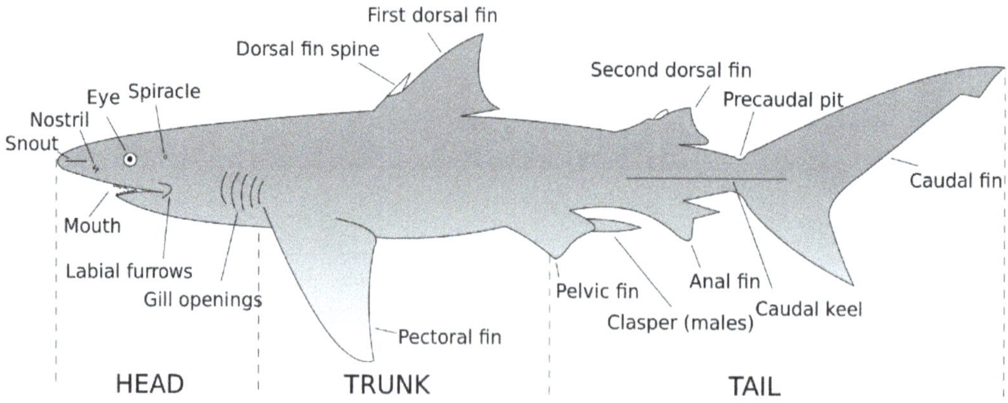

Parts of a shark. photo by Chris_huh **via public domain, March 7th 2007.**

They are shorter than other sharks with a blunt nose that they occasionally use to head butt prey before attacking. They are gray on top and white below and have dark tip fins. This allows them to hunt better because prey below them see white from above masked with the light from the sun and animals above them see gray which can be hard to see blending in with the ocean floor. This is known as countershading. The bull shark is diadromous, meaning they can swim between salt and freshwater with no issues. The reason for this is due to a bottleneck effect where groups of bull sharks were separated during the last ice age. This bottleneck effect allowed bull sharks to adapt to both freshwater and saltwater. Bull sharks blood is naturally salty to allow them to live in saltwater. When moving from saltwater to freshwater where the salt levels are much lower they increase the level of osmosis in their gills to make up for the loss of sodium and chlorine. They also have a number of organs that can regulate the amount of saltwater in their body to allow them to easily live in freshwater, organs like, rectal gland, kidneys, liver, and gills. Although Bull sharks can live in freshwater it may not be advantageous for them to do so, mainly because their main source of prey is located in saltwater areas.

Bull sharks mainly eat bony fish. However, they are not picky eaters. They will eat dolphins, turtles, stingrays, even other Bull sharks. Humans are not necessarily on their menus but they have been known to bite humans out of confusion for other prey or curiosity. They use a technique called the bump and bite attack. After the first initial contact, they continue to bite and tackle prey until they are unable to flee.

Bull sharks mate during the late summer and autumn seasons. They can have anywhere from 1-13 live young after gestating for 12 months. Gestation is the process of the development of a fetus inside the body of viviparous animals.

A Caribbean reef shark **photographed at Roatan, Honduras.**

20 Chapter | LONGFIN DAMSELFISH

Coral reefs are home to more than 35,000 marine species. These species range from simple plankton all the way up to complex reef sharks. One little fish that I have found to be extremely interesting is the Longfin Damselfish. Every living organism in the world has a common name and a scientific name. The scientific name of Longfin Damselfish is *Stegates diencaeus* in the family *Pomacentridae.*

Adult Longfin Damselfish reach a length of 12.5 cm when they become full grown. Adults are commonly a blackish-grey color and their snout and nape of the neck have a yellowish-brown tint. On their anal fin, you can see a bright blue coloring just on the outer most part of their fin. Their head is generally small and contains a small row of sharp front teeth. Their dorsal fin is singular and continuous with 12 spines and 14-17 rays. Their anal fin also contains rays and 2 spines. Just like humans, juveniles and adults look much different. When Longfin Damselfish are younger, their bodies are bright yellow. There are two purple-blue stripes that start at their head and continue all the way back to their anal fins. Once the stripe gets to the anal fin, it ends with a large black dot just at the base of their anal fin.

Longfin Damselfish juvenile by zsispeo **via Flickr**

Longfin Damselfish by zsispeo **via Flickr**

Longfin Damselfish can be found in the Atlantic Ocean, mainly in southern Florida, Bahamas, Mexico and the Carribean. They like to make their homes in coral and rocky reefs. They are a very territorial species and will spend most of their lives by themselves or with their mate. It is unknown at what points during the year Longfin Damselfish mate, but, we do know that they always spawn at dawn. When the mating season begins, the male and female engage in a "mating dance" with rapid swimming and fin movements. During this time, the males will turn a shade or two darker and some display white blotches. Once the female picks her mate, she lays a ton of sticky eggs that stick to the nest. The male then comes and fertilizes the eggs. After fertilization, the male will protect the eggs until they are hatched. This usually takes roughly 2-3 days.

Longfin Damselfish consume mostly algae, plankton, and benthic invertebrates. This puts them in the category of secondary feeders. The only natural known predator that they have is the Lion Fish. While this predator is an invasive species with no known predators of their own, the Longfin Damselfish species is nowhere near endangered. Longfin Damselfish are also caught for the aquarium trade. Fishermen have to use nets and traps to catch the Longfin damselfish. They are known as nibblers so they can not be caught with hook and line. While they are caught by humans for sale, it does not appear to be globally affecting their populations.

20 Chapter | PARROTFISH

Parrotfish are colorful fish found throughout the world's coral reefs. There are about 90 species of parrotfish, including Stoplight, Queen, and Princess Parrotfish common in tropical and subtropical parts of the western Atlantic Ocean and the Caribbean Sea. Parrotfish are characterized by a beak-like structure (fused teeth) that they use to feed. They are mostly herbivores and feed on algae, but also feed on corals and sponges. Their beak helps them scrape algae and crush the hard limestone of corals. They then excrete this undigested material as sand, which helps to create the white sandy beaches of the tropics.

Parrotfish are also known as sequential hermaphrodites. A sequential hermaphrodite is an animal that goes through an initial and terminal phase during its adult life. Each phase is characterized by being one gender. So, for example, some species of parrotfish start as female in their initial phase, then change to male in the terminal phase. This change usually takes place due to an environmental cue, such as the loss of the dominant male.

StParrotfishoplight parrotfish, terminal phase (CC SA 3.0)

Probably the most interesting thing though about parrotfish is their role in coral reefs. Coral reefs face a number of stressors including ocean acidification, rising ocean temperatures, and changing ecosystem balance due to overfishing and bycatch. Interestingly, studies show that the **number one** thing we can do to protect the health of coral reefs is to limit the number of parrotfish we remove from the environment. In the Caribbean, parrotfishes are the primary herbivores on the reef at mid-depth, helping keep macroalgae in check. A shift to macroalgae dominated habitat would offer little value to fisheries, as most of the nutrients are lost to detrital pathways. Unfortunately, parrotfish are commonly fished in the Caribbean. A recent study suggests that if we implement a capture size restriction of less than 30cm there is a win: win outcome in the short term. This would have both ecological and economic benefits as it would also lead to an increase in coral reef health and production. Focusing on more long term benefits requires a more strict harvest limitation to combat the ever-increasing threats to coral reefs.

Parrotfish are also one of the primary grazers of sponges. Sponges (along with corals) are the primary habitat-forming organism on Caribbean reefs. Loh and Pawlik found that parrotfish, along with other spongivores, would graze on sponges that lacked chemical defenses over sponges that possess secondary metabolites. Due to this grazing, the palatable sponges tend to heal and grow faster, as well as have a higher rate of recruitment and reproduction. This allows them to compete with sponges that are left relatively untouched. They also determined that sponge species composition depended more on the abundance of spongivores instead of geographic location. A decrease in the number of these species would result in a top-down effect, leaving more of the faster spreading palatable sponges to out-compete the slower defended sponges and reef-building corals, worsening the state of these coral reef communities. Below is a figure from this paper which compares different sites and the percent of sponge communities and how it relates to an abundance of spongivores.

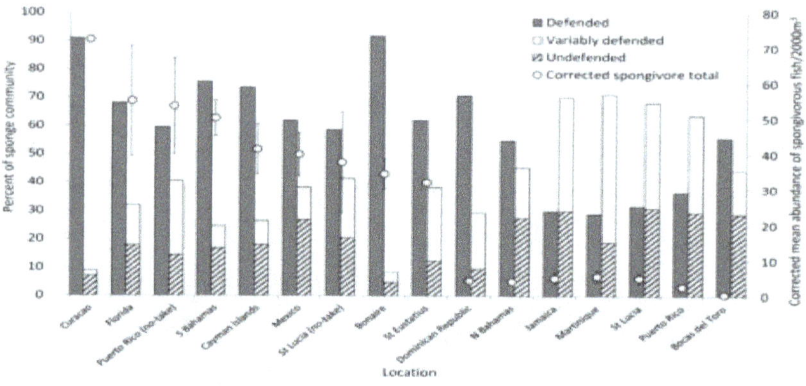

Figure 2 from the Loh and Pawlik paper

22 Chapter INDIGO HAMLET

Hypoplectrus indigo, or the Indigo Hamlet as it is commonly known, is in the order Percoidea and the family Serranidae. Indigo Hamlets typically grow to be 3-5 inches (about 8-13 centimeters) in length and are known for their beautiful blue and white lateral "bar" striped bodies. They typically have blue colored ventral fins and opaque clear/white pectoral fins and caudal. The fin lines and dorsal fin range from blue to opaque clear/white in color, and they tend to mate with other Hamlets of similar color and stripe patterns.

Indigo Hamlet **by** Tomh009 **via Flickr**

Although the patterns of striping can differ a bit, all Indigo Hamlets possess the similar blue and white diagonal 5 stripe pattern throughout their life phases which are spent within the safety of the reef. Hamlets are small, oval-shaped fish that have sloping heads and tapered bodies. The Indigo Hamlet ranges throughout the Caribbean island reefs, from the Bay Islands to the Bahamas, Belize, the Cayman

Islands. Hamlets are known to stay near the benthic region on and around the coral reefs between 3 and 45 meters, however, that is not to say that they are inactive, as they are often seen swimming around. They are relatively solitary fishes who tend to stick to themselves rather than swim in large schools and are described as highly resilient to the environment. Although they themselves are not a protected species, they do live in some protected reef areas. Hamlets are also simultaneous hermaphrodites, meaning that they have both male and female sexual organs and are able to act as both genders at the same time, which is incredibly rare for vertebrates. Although they do not self fertilize, they do take turns mating as both male and female.

Indigo Hamlet 1 – Blackbird Caye – Belize 2016 **by** Adam **via Flickr**

The Hamlet is a carnivorous fish who feeds predominantly on other species of fish and is relatively territorial within its habitat. Because of this, it is possible that the Indigo Hamlet possesses its beautiful coloration for a number of reasons that are similar to other fish species. The patterns and coloration may announce territorial ownership to others, secondly, it may help with courting, and third, it is possible that the bright coloration would be useful for protecting their eggs and hunting grounds. By being brightly colored and patterned, mates will easily spot them as their own species. Furthermore, their bright color and aggressive behavior masssssw draw others away from their territory and eggs.

23 Chapter | PLASTICS IN OUR OCEANS

Plastics are an integral part of our modern lives. They are found everywhere from Home Goods to electronics, and are by far one of the most versatile and useful technologies humans have been able to harness. On top of their usefulness, plastics are easily re-purposed, with more than 40% of plastic being recycled. Although plastic helps us go about our everyday lives, it is having a serious effect on the ocean and marine life. In 1975 it was reported by the National Academy of Sciences that an estimated 14 billion pounds of garbage, much of it plastic, were being dumped into the oceans every year. Today, that number has shrunk to around 8 billion pounds a year thanks to public awareness.

SO WHAT ARE THE EFFECTS PLASTIC HAS ON MARINE LIFE?

For starters, "...Plastic has been found in more than 60% of all seabirds and in 100% of sea turtle species..." according to the group Ocean Conservancy To a sea turtle, a plastic bag can look extremely similar to a jellyfish, which is a common food source for them. For other species such as seabirds, fish are a common food source. When smaller pieces of plastic get consumed by small fish, the plastic will move its way up the food chain to bigger fish and bigger fish until it is consumed by a seabird, or other marine animal. The accumulation of plastic in these species impacts their health in many ways including nutrient uptake, general fitness, and feeding efficiency.

"Contrasts of Concern" by Leonard J Matthews [CC by 2.0]

While fish and turtles are motile animals, sessile species such as corals are not free from the effects of plastics in their environment either. A study has been done by Hall et al. that shows corals will consume micro-plastics at the same rate as their food. Micro-plastics are any piece of plastic under 5mm in size. The corals in this study were found to starve at a very slow rate because the plastic pieces that they were consuming overloaded their stomach and they were not getting the amount of nutrients that they needed. Corals are very non-selective feeders, but they are sensitive to their environment.

Although the oceans are huge, they are not big enough to disperse all of the plastic that winds up in them. It is estimated that there are between 15-51 trillion pieces of plastic in the worlds oceans, and due to ocean currents, millions of tons of plastics congregate in massive gyres in the ocean. There are 5 major subtropical gyres, each containing its own garbage patch. The most notorious of these is the Great Pacific Garbage Patch, an island of plastic waste 1.6 million kilometers squared in area, or three times the size of France. In this garbage patch, it is estimated that there is 6 times more plastic than plankton. Plankton is a necessary food source for some marine life. It is also up to 9 feet deep in some portions of the garbage patch. For more information on garbage patches, watch the video below:

"Gyre" by Chris Jones via Vimeo

One of the questions this raises is how does it all get there? How does 8 million tons of plastic waste end up in the ocean each year? In the developed world, which contributes for roughly 10% of the worlds plastics problem, plastic enters our oceans through our sewer systems, and as consequences of storms and heavy rains. While the developed world contributes in part to the plastic problem, we are by no means the biggest culprit. According to the World Economic Forum, 90% of all plastic waste that pollutes our oceans comes directly from 10 rivers in China,

Southeast Asia (Laos, Thailand, Vietnam, etc), Egypt, and Niger. What these areas have in common are massive populations centered around these rivers combined with poor waste management systems and general lack of education about environmental preservation.

"Garbage in the River Ganges in Varanasi" by Brett Cole Photography [CC by 4.0]

So what is being done to combat this?

There have been a number of laws put in place in an attempt to stop the plastic pollution. The Act to Prevent Pollution from Ships (APPS) (PDF) was amended in 1987 by the Marine Plastic Pollution Research and Control Act. It is meant to reduce the amount of improperly disposed plastic that is released into the environment. It also studies the effects that this plastic is having on the environment.

The Marine Debris Research, Prevention, and Reduction Act (MDRPRA) (PDF) has established programs within the National Oceanic and Atmospheric Administration (NOAA) and the United States Coast Guard. Its goal is also to reduce the amount of plastic and trash that gets put into the marine environment.

The Shore Protection Act (SPA) is another Act that was put into place in order to stop this growing problem. This Act controls the transportation of waste in coastal waters.

One of the most important Acts put into place is the Marine Protection, Research, and Sanctuaries Act (MPRSA). It is also known as the Ocean Dumping Act and it prohibits the following:

- Transporting material from the United States with the intention of ocean dumping
- Transporting material from anywhere for the purpose of ocean dumping by U.S agencies or U.S flagged ships
- The dumping of material transported from outside the U.S into the U.S. territorial sea

Outside of the US, thanks to efforts by international and local organizations, agreements have been signed and efforts are being made to reduce plastic wastes across the globe. China for example has ordered 46 cities to recycle 35% of all plastics by the year 2020, and India has begun to ban disposable plastics in some cities.

24 Chapter | NOISY OCEANS

Human noise in the ocean is becoming a major concern. From sonar on our submarines to cargo ship traffic and even oil drilling rigs, we are filling our oceans up with lots of unnatural noise. This noise is linked to altering the calling, foraging, and migration patterns for many species. In extreme situations, scientists also believe it is a reason there have been so many beached whales.

Here is a video that shows what some different noises in the oceans sound like, and how loud they really are.

"Virginia Class Submarine" by Owly K Under U.S. Federal Government

EFFECTS OF SOUND ON WHALES

It seems that whales and dolphins are the marine animals that are most affected by the noisy oceans. Although natural noise in the ocean from wind, waves or other marine animals is a common occurrence, they are not nearly as intense as noise from humans. This more intense and frequent noise can be causing stress to whales and contributing to them beaching themselves. One scientist has said that he has seen whales abandon a location because of noise. These locations sometimes contain a food source necessary for the whales survival.

Sound waves from military submarine sonar systems can get as loud as 235 decibels. They are able to travel hundreds of miles and can maintain the intensity of 140 decibels as far as 300 miles from the source.

Cargo Ships create what is known as ambient noise. This is particularly concerning for low-frequency great whales. Right whales are endangered and it has been proven that noise from cargo ships constantly crossing the ocean is causing them stress.

"Cargo Ship" by Peter Griffin Under Public Domain

Here is a video on the process of deep water oil rigging. It would be hard to believe that these rigs don't make a ton of noise. It is understandable that hearing noise from these rigs constantly could cause stress to a whale or confuse it considering they rely heavily on acoustics for survival.

RECENT BEACHED WHALE STORIES

On February 10th, 2017, New Zealand had over 600 Pilot whales wash ashore. Volunteers were able to save about 80 of the whales, who went on to join a nearby pod. After they were saved, later that night the pod that they joined got stranded on Farewell Spit. Although the cause of these whales stranding and beaching themselves is unknown, one theory is underwater noise. The loud sound waves that humans create can panic the whales into surfacing too quickly or swimming into the shallows.

“Whales on Beach, Farewell Split, South Island New Zealand” by Chagai Under Public Domain

In May 2016, more than 20 whales beached themselves near San Felipe, Mexico. According to one statement, there were no signs of injuries noted on the whales, but they seemed disoriented. Since the whales had no signs of injury, it is very reasonable to believe that sonar and loud ocean noise drove them out of deep water towards the beach.

Less than a year after 330 whales washed ashore on Patagonian Inlet, 70 whales were found on a beach in southern Chile. Although they are not the same species, this is a terrible trend that is happening. The whales were smaller this time and they had been dead for about 2 months before they were found.

25 Chapter
NOISY OCEANS

Ocean acidification is simply the decrease in the pH of all of the oceans on Earth. But is it really that simple? The answer is no, it is not that simple. The ocean absorbs about 25% of the atmosphere's carbon dioxide or CO_2, and as atmospheric CO_2 levels increase, so does the amount of CO_2 that the ocean absorbs. Therefore, as the amount of greenhouse gases increase, not only does the temperature of the ocean begin to rise, but the increase in CO_2 levels causes the ocean's pH levels to decrease. The rising acidic conditions cause a multitude of deleterious effects from limiting the formation of skeletons for marine organisms, to limiting coral growth, and corroding already existing coral skeletons.

While natural levels of carbon dioxide are fine, the excess CO_2 that humans have managed to produce through burning fuels are the major problem. This is largely because when CO_2 dissolves in the sea water, ($CO_2 + H_2O \rightarrow H_2CO_3$), it produces carbonic acid. While carbonic acid is not as strong of an acid, as say HCl, it still acts as an acid by donating protons, and interacting with surrounding molecules. Now, when excess acid is added to a solution, it causes the pH of that solution to drop, or become more acidic, and while H_2CO_3 is not considered a strong acid, it still has the potential to massively impact the entire chemical makeup of the ocean environment. The lower the pH of a solution gets, the higher the concentration of H^+ ions in that solution. This has become a massive problem for many species.

Chemical reactions can be extremely sensitive to any fluctuation in pH levels, however, in the ocean, these pH changes can affect marine life through chemical communication, reproduction, as well as growth. In particular, the building of the skeletons in marine life is extremely sensitive to any change in pH. Thus, the current increase in CO_2 levels that have caused a more acidic environment, have greatly and negatively impacted the growth of new shells.

"Nautilus pompilius" by Wikimedia [CC BY SA 2.0]

This is because Hydrogen ions easily bond to carbonate (CO3 2-) molecules to create carbonic acid, and a marine animal's shell consists of Calcium Carbonate (CaCO3). In order for marine animals to build shells, they take Calcium ions (Ca2+) with a carbonate molecule from the seawater surrounding it, to form the Calcium Carbonate needed to build their shells. So, instead of the carbonate giving all of it's attention to Calcium so that the marine animal can build it's shell, it now pays more attention to the Hydrogen ion. In addition, Hydrogen ions have a greater attraction to carbonate, than a Calcium ion does to carbonate, and when two Hydrogen ions bind to carbonate, it produces a bicarbonate ion, and a marine animal does not have the capability to extract only the carbonate ion. Ultimately, this limits marine animals from building any new shells for themselves, and even if the marine animal has the ability to build a new shell, it takes a lot of energy to do so, essentially taking away from other important processes and activities.

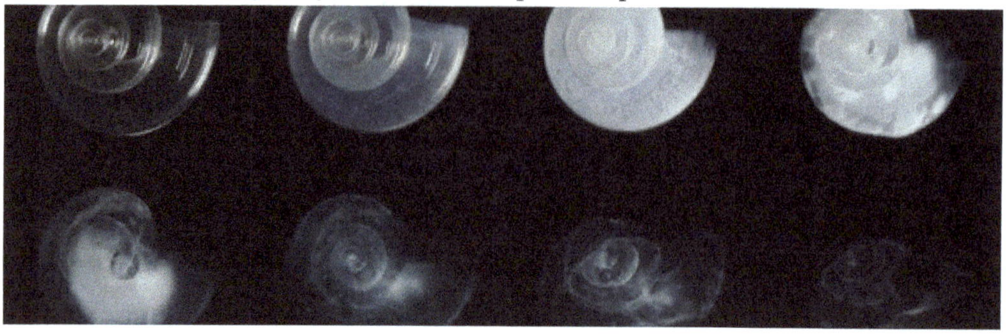

"Impacts of Ocean Acidification" by Wikimedia [CC BY SA 2.0]

Not only can this affect the formation of new shells, but in the right conditions, it can corrode already existing shells. When there are too many Hydrogen ions floating around, and not enough molecules for them to bind to, they can actually start breaking the already existing Calcium Carbonate molecules apart, ultimately breaking down the marine animal's shell that already exists.

Another organism that feels the effects of ocean acidification are corals. Similar to other marine animals that build their homes with calcium carbonate, reef-building corals also use calcium carbonate to build their own bodies and structures. These reef-building corals are also home to other coral animals and other organisms. As mentioned before, the increased acidification greatly limits any further growth of new coral, as well as corrodes any pre-existing coral reefs. Even if a coral reef has surpassed all the odds, and has been able to grow, it will be a weaker reef that is subject to natural erosion. It has been predicted that by the middle of the century, healthy coral reefs will be eroding more quickly than can be reproduced.

"Blue Starfish on Hard Coral" by Wikimedia [CC by 3.0]

This is bad news for the species that live in these coral reefs. If these species are not able to grow and develop in a safe settlement, such as the coral reef, these larvae will be unable to reach adulthood, thus being unable to reproduce. This will ultimately lead to a mass extinction in the future.

25 Chapter

HARMFUL ALGAL BLOOMS

Ever been told to only eat shellfish during the months that have the letter "R", (September-April)? Well this rule is actually pretty important for keeping the health of people safe and to allow for many species of shellfish to repopulate. But why are the other months of the year not safe for people to eat shellfish? In short its because of the algae that grow during this time of year and as ocean temperatures rise. During these specific months of warmer weather, billions upon billions of these microorganisms start to take over our oceans and can have many consequences for us.

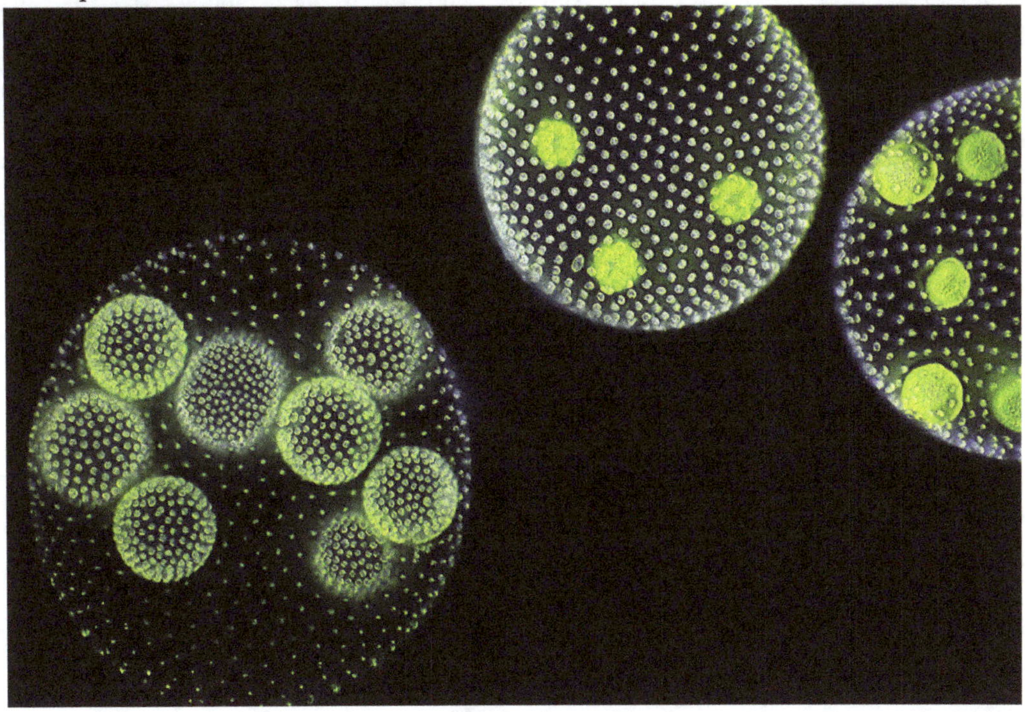

"Multicellular Green Algae" by Frank Fox [CC by 2.0]

Before going into what red tide is or how the populations of these microorganisms seem to be increasing significantly as oceans warm up, lets take a closer look at algae. Most species of algae are single-celled organisms but some species can be multi-cellular as seen in the photo above. Algae are autotrophs, meaning they use photosynthesis as their means of producing energy for themselves. Though similar to plants in the way they are both producers, algae have no stems or leaves and are more closely related to other groups of protists. Habitats for algae include any bodies of water including fresh and salt water, or have extreme external environment factors. There are few cases where they have been found on land such as rocks, trees, hot springs, etc... Species of algae have been well documented to be able to survive many harsh environments and have been on earth far longer than most living organisms to this day. They contributed to the Earth being able to house life by producing oxygen through photosynthesis. Overall Algae species are very tough and can survive in a wide range of environments, which can be seen as both a positive and negative situation.

The red tide occurs when the algae from algal blooms becomes so numerous that it discolors the water. It is also sometime referred to as a Harmful Algal Bloom or "HAB". This is where the name "red tide"comes from. Some key factors involved in red tides forming are warm ocean surface temperatures, low salinity, high nutrient content, calm seas, and rain followed by sunny days during the summer months. Some effects of the red tide are that it could deplete the oxygen in the water and/or release toxins into the water. The toxins in the water could have negative effects on the health of humans and animals exposed to them. There are three types of algae that can release these harmful toxins, they are *Alexandrium fundyense*, *Alexandrium catenella* and *Karenia brevis*.

"Red Tides in Isahaya Bay, Japan / Algal Bloom in Pelee Island, Ontario" by Marufish and Tom Archer [CC by 2.0]

What is important to recognize about "Red tides" and Algal blooms is that it isn't always obvious that algae growth is there. They are not always a red color. The photos above show two examples of Algal blooms from two very different parts of the world, yet both species are considered "Red Tide" and harmful to some shellfish and animals that eat the shellfish. Algae alone is not an issue and even during the time of the year where there seems to be an excess amount of growth, this is a natural occurrence. What becomes a problem or what classifies as a "Red Tide" are the algae that release toxins in the air and water when they grow. Very few algae species can produce this toxin but when a large enough group forms on shores it can have a negative effect on both the marine environment and humans. The toxins produced can often affect the respiratory and nervous systems of all life forms. Thus when smaller marine animals feed on the algae, the trophic level above them can become poisoned as well. Paralytic Shellfish Poisoning is typically found along the Pacific and Atlantic coasts of the United States and Canada. It can cause paralysis and in extreme cases death. Some of the toxins that cause Paralytic Shellfish Poisoning can be 1,000 times more potent than cyanide. Diarrhetic Shellfish Poisoning is another example of a harmful effect from eating contaminated shellfish. It is caused by Okadaic acid, which is produced by several species of dinoflagellates, and is usually non-deadly to humans. Small amounts of the okadaic acid usually do not have any harmful effects and only become an issue when large amounts are consumed. Amnesic Shellfish Poisoning is the third common poisoning that humans will get from eating contaminated shellfish. It can be life threatening and cause both gastrointestinal and neurological disorders. These disorders are caused by domoic acid. After an incident in Canada in 1987 where 4 people died from Amnesic Shellfish Poisoning, the levels of domoic acid in shellfish are now being monitored.

"Toxic Algae Bloom in Lake Erie" by NASA Earth Observatory Under Public Domain [CC by 2.0].

Algal Blooms can have serious effects on corals. Red algae, brown algae, and green algae are a few examples of macro-algae that can have a very negative effect on corals. They do this by outcompeting, overgrowing and eventually replacing sea-grasses and coral reef habitats. According to some research that is being done, harmful tropical algal blooms are increasing in frequency and intensity. This can have a significant impact on coral reefs.

NOTABLE RED TIDES

- 1844: First recorded case off the Florida Gulf Coast.
- 1972: Red tides killed 3 children and hospitalized 20 adults in Papua New Guinea.
- 2005: The Canadian red tide was discovered to have come further south than it has in years prior by the ship (R/V) Oceanus, closing shellfish beds in Maine and Massachusetts. Authorities were also alerted as far south as Montauk to check their beds. The experts who discovered the reproductive cysts in the seabed warned of a possible spread to Long Island in the future. This halted the area's fishing and shellfish industry.
- 2013: In January, a red tide occurred on the West Coast Sea of Sabah in the Malaysian Borneo. There were two fatalities reported after they consumed a shellfish that had been contaminated with the red tide toxin.
- 2015: In September, a red tide bloom occurred in the Gulf of Mexico, affecting Padre Island National Seashore along North Padre Island and South Padre Island in Texas.

"Map of Padre Island National Seashore" By U.S. National Park Service Under Public Domain [CC 0]

Scientists have been able to help control the spread of the effects of harmful algal blooms by developing new technology to help track them better. Tracking these harmful algal blooms could help prevent people from eating contaminated shellfish and knowing which areas will be most effected by them. Some examples of the technology that can help with monitoring them are better and more advanced satellite imagery. Also the development of an antidote to the toxins produced is another way to reduce the harmful effects. Even though they are a natural occurrence, what is alarming some scientists is that they may start to last longer and occur more often during the year as ocean temperatures and CO2 levels rise.

"Red Tide in Naples" [CC by 2.0]

27 Chapter | CORAL BLEACHING

"Eco Cafe" by Phuket [CC by 2.0]

Coral bleaching is a phenomenon that occurs when coral turns white due to environmental stress. It is a reoccurring event that frightens marine biologists and raises a lot of concern for marine ecosystems all over the world. When the water is too warm, corals will expel the algae (zooxanthellae) living in their tissues. As a result, the coral loses its vibrant color and becomes more prone to developing disease. Corals have a mutualistic relationship with these algae. This means that the organisms depend on each other for survival. From their photosynthetic products, zooxanthellae give the coral the nutrients they need to survive while the coral in exchange give the zooxanthellae carbon dioxide and safe harbor. When they are lost, it is a huge loss for the coral because these symbiotic algae provide the coral with 90% of its energy through the process of photosynthesis. Without the zooxanthellae, the coral begins to lose its color and eventually has a bright white appearance, hence the term coral bleaching. In essence, the coral will starve and die without the needed nutrients from the zooxanthellae. Global warming is a big cause of coral bleaching but other environmental factors also play major roles.

The video below explains more about what coral bleaching is, what causes it and the effects that it is having!

The corals may be able to survive for a short period of time without the zooxanthellae, but since the algae provides its primary food source it cannot survive for long. Once the coral has died, its begins to degrade. Another cause of coral bleaching is oxygen starvation. This is caused by an increase in the zooplankton population in a given area, most likely due to over-fishing. Another trigger that is less common but still notable is sedimentation. This occurs from silt runoff from land into the ocean via rain and other bodies of water leading into the ocean. Also, La Nina or EL Nino can also play a role by not only changing the temperature of the water but also by bringing some silt/sand along with nutrients up from the bottom of the ocean. The biggest cause of coral bleaching is from the stress created from warmer water temperatures due to global climate change.

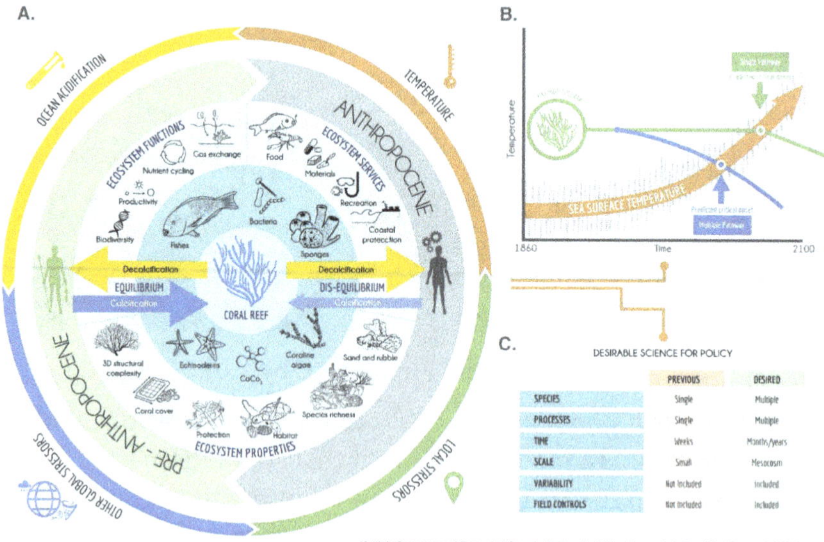

Multiple Stressors and Ecological Complexity Require A New Approach to Coral Reef Research (2016).
Pendleton Linwood Hagan, Hoegh-Guldberg Ove, Langdon Chris, Comte Adrien. Frontiers in Marine Science 3, 00036.

“Ecological Processes Coral Reefs” by Wikimedia [CC by 2.0]

Corals provide shelter for a number of marine organisms and they also protect shorelines. Without coral reefs, marine life would most likely leave the area which in turn, could cause harm to marine ecosystems. Coral reefs help to support approximately 25% of marine species. They are also a huge benefit to the fishing industry since they attract so much marine life.

"Coral Reef Story Intro Video" by GRID-Arendal [CC by 2.0]

There have been three major coral bleaching events recorded in history, the first occurred in 1998, an underwater heatwave spread, killing 16% of corals around the world. The second major coral bleaching event occurred in 2010, caused by an El Niño. The most recent coral bleaching event occurred in 2015, which was caused by ocean warming.

The warming of the ocean is a huge threat to coral reefs. As stated previously, ocean warming is the number one cause of coral bleaching. A shocking 93% of climate change heat is absorbed by the ocean.

BUT WHAT IS THE UNDERLYING CAUSE OF ALL OF THIS?

Global climate change is caused by increased greenhouse gases, primarily carbon dioxide, that are released into Earth's atmosphere. The carbon dioxide and other gases trap heat in the atmosphere, causing the Earth to heat up. When the Earth heats up, water temperatures increase and result in coral bleaching.

Humans are the main source of greenhouse gases that are produced and released into the atmosphere due to the burning of fossil fuels. Vehicles are one of the many producers of Carbon Dioxide. Burning one gallon of gas produces approximately 24 pounds of Carbon Dioxide. Another way in which humans produce these gases is by burning coal for electricity. Burning wood to produce fire for cooking is also another every day activity that contributes to the emission of these gases. The other major factor that contributes to increase in greenhouse gases in Earth's atmosphere is deforestation. For years, humans have been cutting down trees in large numbers to make room for the construction of buildings and also to use the wood as fuel. Trees are very important to our ecosystems because they absorb a lot of the carbon dioxide that is in the Earth's atmosphere. By absorbing carbon dioxide, they are able to perform photosynthesis and release Oxygen as a byproduct. As trees are removed in large numbers, this sink for carbon dioxide is greatly reduced and the amount of carbon dioxide in the atmosphere continues to increase.

"Blackback Butterflyfish" by Wikimedia [CC by 2.0]

Most people don't realize how important coral really is to not only marine life, but also to human life. For fish communities and other reef dwelling creatures, the coral is everything to them and they all rely on the coral structure in some way. The reefs provide shelter for smaller fish and can be a food source for others. Coral reefs are essential for life for a countless number of organisms and without the coral, a good number of these organisms will perish, leading to a much bigger problem. If the smaller fish can no longer survive without the coral reef, then that will lead to less food for the larger fish that feed on them.

Around 100 million people around the globe rely on coral reefs for survival. With a decrease in healthy coral ecosystems, there will be less fish which are an important food source for many humans. People also rely on coral reefs for other economic purposes such as tourism. The impact of coral reef loss on the tourism industry is estimated at around 10 to 40 billion dollars, followed by fisheries losing 7 to 23 billion dollars and biodiversity impacts resulting in losses of 6 to 22 billion dollars.

MARINE SPECIES IMPACTS:

"Polyp Butterflyfish" by Bernd [CC by 2.0]

Butterfly fish are fish that feed exclusively on coral polyps. With their food source disappearing. their chance of survival is small. Butterfly fish also keep algae from smothering corals and if this population decreases, the corals that are still living could end up dying because of too much algae.

"Spiny Lobster Approaching" by Eric Kilby [CC by 2.0]

Spiny lobsters need coral reefs for protection, especially when they are molting because that is when they are most vulnerable. These lobsters play a significant role in maintaining a balanced ecosystem. Spiny lobsters are predators of sea urchins, which feed on kelp forests and can destroy them if populations are not predator-controlled.

"Dolphin" by Wikimedia [CC by 2.0]

Dolphins and most whales are predatory animals and they are all carnivorous. They survive by eating fish that rely on coral reef habitats. Each species that is lost from coral bleaching has the potential to affect the rest of the ecosystem.

"Hawksbill sea turtle" by Raymond™ [CC by 2.0]

Hawksbill Sea turtles have become a critically endangered species and are very dependent on coral reefs for their food sources. Their diet mostly consists of sponges. Sea turtles play a very important role in their ecosystems by helping with nutrient cycling from ocean to land, maintaining healthy sea grass beds, and balancing food webs.

"Whale Chark" by Zac Wolf [CC by 2.5]

Studies have shown that whale shark populations have decreased in the 1980s and 1990s and this may be due to the destruction of corals. Many corals were destroyed in this time period from people and coral bleaching. Although it is not certain that these two are related to each other, the times frames suggest that they are.

"Healthy Corals" by Wikimedia Commons [CC by 4.0]

Solutions

There are many ways that people can help reduce the amount of coral bleaching that is occurring. One way that we can do it is by reducing fuel pollution by using alternative transportation methods such as running or biking instead of driving a motor vehicle, or kayaking and canoeing instead of using your boat. Not touching and breaking corals when diving or snorkeling can be a huge help since just touching a coral with your bare hand can have negative effects on it. Not using harmful fertilizers and pesticides in your yard and garden can also help a lot because they will eventually find their way to the ocean. Volunteering for a community reef and beach cleanup can also help keep trash out of the ocean which will keep it from getting into the reefs.

28 Chapter THE WARMING OF OUR OCEANS

When talking about the warming of our oceans, it's important to first talk about global warming. Is global warming real?

This topic has been debated for years and even more so now with the new United States president. This topic has sparked a great debate among caring citizens, politicians, and news outlets on whether or not global warming is real or a hoax. However, scientists all over the world have scoured over all relevant data and facts and harmoniously agree the planet is indeed warming.

Many might ask, "well what exactly is global warming?". Global warming is the term used to describe a gradual increase in the average temperature of the Earth's atmosphere and its oceans; a change that is believed to be permanently changing the Earth's climate.

There are cold hard facts that prove climate change is prevalent. The EPA has over forty data contributors from different government agencies that provide indicators of the cause and effects of climate change. These indicators effect the oceans in detrimental ways like thermal stress, ("stress in a body or structure due to inequalities of temperature"), and the warming of the oceans.

GREENHOUSE EFFECT

The greenhouse effect is the result of greenhouse gases trapping heat in the Earth's atmosphere radiating from earth toward space, resulting in the warming of the oceans and melting glaciers. When the glaciers melt, it increases the sea level which adds to the ocean that is already expanding.

Examples of greenhouse gases are: carbon dioxide (CO2), methane (CH4), nitrous oxide (N2O), and water vapor (H2O). From 1990-2010 gas emissions caused by humans have increased by 35%. Electricity generation is the worlds largest contributer of emissions followed by transportation. Climate forcing, which is a change in the earth's energy balance that results in a warming or a cooling effect on the climate. Over the past several years we have also steadily increased our use of fossil fuels, which has ultimately produced 5x more carbon dioxide emissions. The graph below shows how our use of fossil fuel emissions has skyrocketed in the last decade.

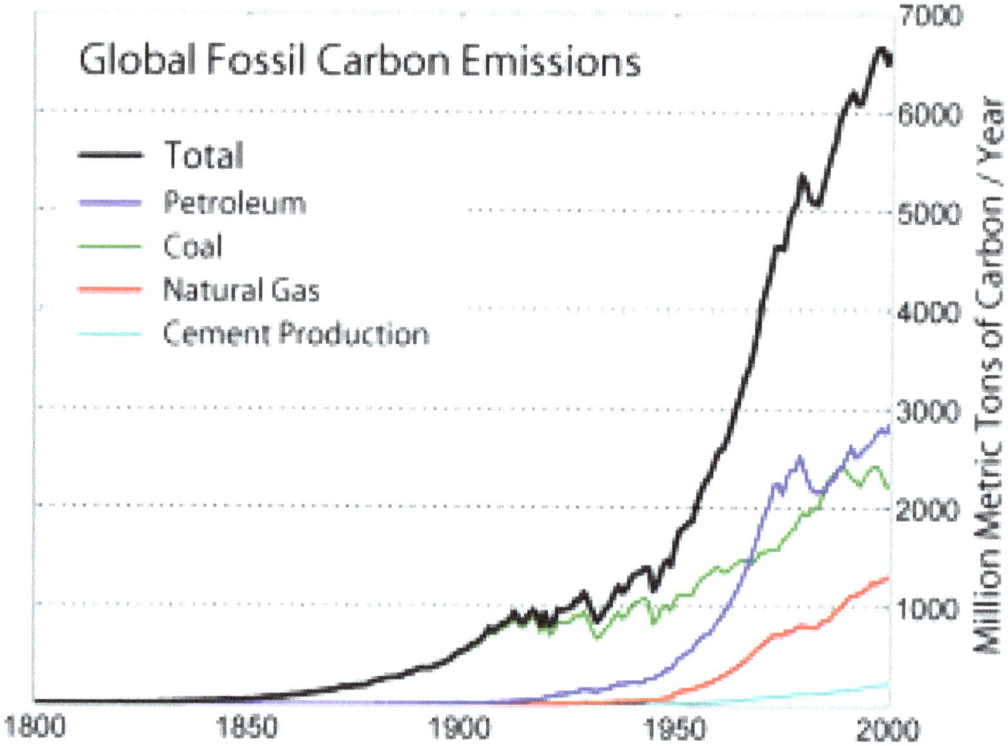

"Global Carbon Emission" by Wikimedia Commons [CC by 2.0]

How is the warming ocean impacting the marine life?

One of the most vulnerable organisms to global warming is coral. Higher temperatures cause coral bleaching, which results in disease, lack of nutrients and even death of the coral. The zooxanthellae's enzyme systems are effected from rising temperatures which makes it hard to protect the coral from toxicity. Even an increase of 1-2 degrees Celsius can cause bleaching. Coral Bleaching occurs when corals experience stress such as temperature change. This harms corals because they have a limited temperature range in which they can live. When water is above the corals ideal temperature, the coral expels the symbiotic algae that reside in its tissue and provide it with nutrients. This turns the reefs a ghostly white (thus the term 'bleached'), and, while the coral is not exactly *dead* at that point, it is more susceptible to diseases which lead to death.

"Coral Bleaching" by Wikipedia [CC by 2.0]

As humans burn fossil fuels and release carbon dioxide, those gases enter the atmosphere where they cause an increase in global temperature. However, did you know that **not** all of the carbon dioxide ends up in the atmosphere? In fact, about 40% is said to get absorbed by the ocean waters. The amount of carbon dioxide that the ocean can hold is dependent upon the temperature of the ocean. For example, colder water can absorb more carbon where warmer water absorbs less. Today, scientists believe that as the oceans warm they will become less and less capable of taking up and absorbing carbon dioxide. As a result, more of our carbon pollution will stay in the atmosphere and therefore further the contribution to global warming. For now, the oceans are considered our saviors by absorbing large amounts of our carbon pollution. This buys us some time to reduce our use of fossil fuels. However, there is always a consequence. The increased carbon uptake by the ocean means that the ocean waters will become acidic more rapidly than they otherwise would. This acidification threatens many components of the food chain.

When the ocean temperature increases, this can cause a bottleneck in the food chain, which can result in an interruption in the marine food web. The higher the temperature, the higher the growth of zooplankton. Zooplankton reproduce faster than the phytoplankton and end up eating all of the plankton in an area. This can have a drastic effect on any organism that feeds off of them.

Rising sea levels are a result of the ocean expanding from warming. When the sea levels rise, they interrupt ecosystems and habitats of thousands of organisms such as sea turtles and seals. A rise in sea level results in coastal flooding, which has increased in the United States on coastlines such as the Carolina's, New Jersey, and Maryland. Floods are now 10 times more likely to happen than they were years ago. These floods are endangering species and ruining coastline habitats.

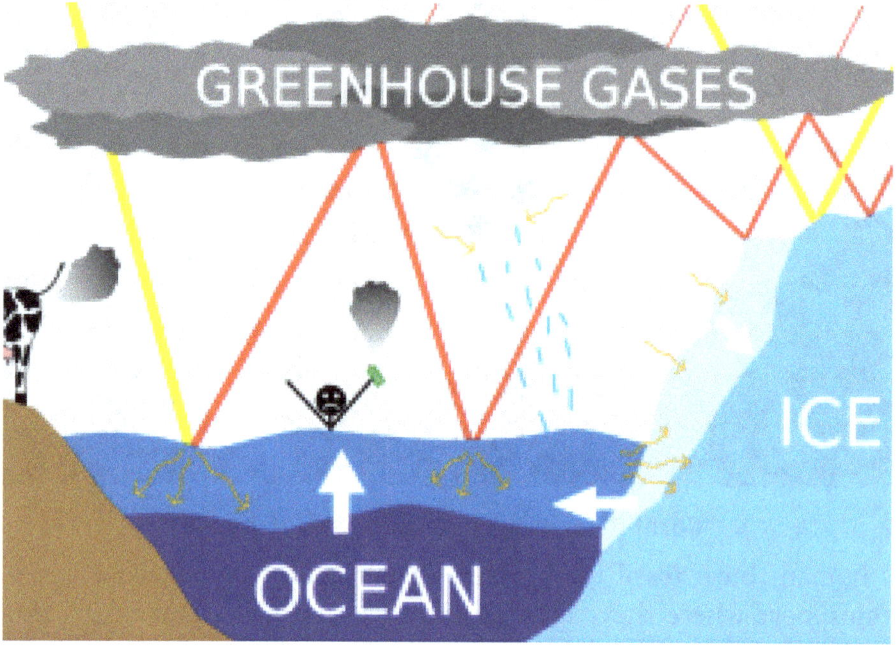

"Sea-Level-Rise" by Wikimedia Commons [CC by 2.0]

Chapter 29 OIL SPILLS AND MARINE LIFE

According to the National Oceanic and Atmospheric Association, there are thousands of oil and chemical spills each year. The Ocean Service's Office of Response and Restoration is the first to come to the scene and assess the impacts that a spill may have, identify risks, and recommend different cleanup methods. However, when oil spills occur, the first organisms to come into contact with the oil are marine organisms. Many of the issues that occur for marine life are due to the bioaccumulation of oil components by organisms. Biological organisms lower in food chains such as zooplankton accumulate the oil's chemicals within their tissues. When these zooplankton are consumed by organisms such as fish at higher trophic levels, the concentration of chemicals is magnified. With each successive trophic level, this biomagnification continues and the concentrations of the chemicals can be extremely dangerous especially for apex predators causing health and reproductive problems.

The effects of oil spills are varied across species, the distance from the spill, how big the spill is, and where the oil disperses. Organisms such as shellfish can be unaffected by oil or only slightly. This is because most of the oil floats within the water column and the amount that sinks to the ocean floor is limited, although there are still some circumstances where the oil spill has a large effect on the shellfish. Oil spills that are in shallow or confined waters are the most at risk for effects. Oxygen depletion can occur due to the formation of oil slicks at the surface of the water.

Oil slicks in deeper water can also have an effect. For example, the BP Oil spill in the Gulf of Mexico had well leaks at deep depths. Organisms such as shellfish that don't move often or far and are filter-feeders, are unable to avoid exposure to oil. Juvenile and adult fish are much more mobile, are more selective in what foods they eat, and they also have a variety of enzymes that allow them to detoxify many oil compounds. As a result, they are often better suited to limited oil exposure and related impacts. In spite of this, many fish are killed as a result of light oils and petroleum in shallow water. Also, oil spills can completely kill or wipe out fish egg populations.

"Louisiana_Oil_Spill" by US Coast Guard, under public domain

"Port Sulphur, La. (Nov. 29)–An aerial view of a section of the Mississippi River containing a dense amount of the Nigerian 'sweet' crude oil spilled by the M/V Westchester Nov. 28, 2000 by USCG" by PA1 Jeff Hall

Effects of oil spills can be direct or indirect. Direct impacts include when oil directly touches, is consumed, or is injected through a cut in the skin. When these things happen they can deteriorate the thermal insulation of some organisms.

They can also result in changes in the behavior and reproductive systems of those organisms that come into contact with the oil. Indirect effects of oil spills are those that result from consuming individuals who have direct contact with the oil as well as effects from the mass mortality and decomposition that occurs during oil spills such as oxygen depletion. Another indirect effect can be losing a major food source which could result in the death or extinction of one or more species.

Another factor that can affect marine organisms is the type of oil that is spilled. Light oils and petroleum products can cause acute toxicity in fish, but the toxic event is generally over fairly quickly. Heavier oils sometimes do not affect fish, however they can be detrimental to fish that are in the larval and spawning stages. The type of oil and the timing of the release influences the severity of oil effects on fish. Heavier oils can have great impacts on sea birds. This is because the feathers on birds are naturally waterproof, and in order to maintain this, the feathers on the birds bodies must be aligned. This is so the water cannot leak through the microscopic barbs, and barbules that are part of the vane of each feather. The bird, through a process referred to as preening, distributes natural oils on the feathers to keep the feathers in place. The oil floating in the ocean water sticks onto the birds feathers, causing it to become matted. The matting causes the feathers to separate ultimately making the feathers no longer waterproof. The bird then suffers from hypothermia or hyperthermia when it can no longer protect itself from extreme temperatures. Birds react to the presence of oil by preening, and by doing so the birds end up ingesting the oil that is toxic to them. During this time all of the birds energy is put into preening and they are left vulnerable and malnourished. Death, in most cases, is what the birds are facing without proper treatment. The washing treatment can not begin until the bird is at an acceptable weight, with good blood values while displaying active and alert behavior.

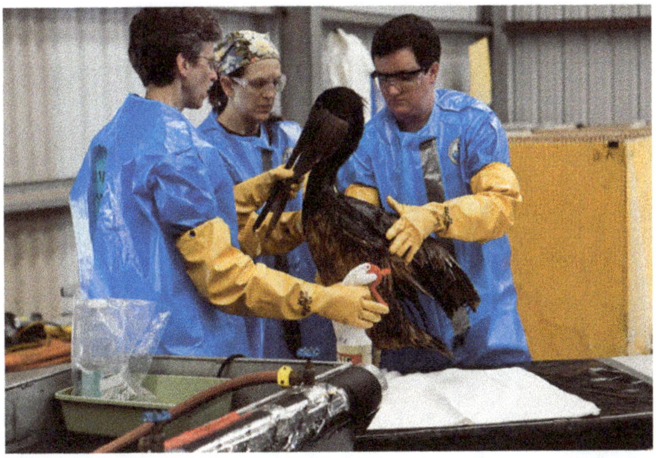

"Oiled Pelican 03 Dawn IBRRC 2010" by International Bird Rescue Research Center [CC by 2.0]

30 Chapter

DESTRUCTIVE FISHING METHODS

When we think about destroying the oceans, most people think of destruction via pollution. Although chemical dumping and plastics have a drastic effect on marine life and ecosystems, it is very important to consider another major issue that is causing detrimental effects on marine systems. This issue is destructive fishing methods. Destructive fishing includes practices that leave marine populations irreversibly damaged and can destroy entire habitats for fish and other organisms.

According to the World Wildlife Fund (WWF), there will be no more fish left in the oceans by 2048. This is because more than 30 percent of the world's fisheries have been pushed beyond their biological limits. Therefore, we need stricter regulations and laws to prevent this disaster.

Destructive fishing is mostly done in underdeveloped countries which don't have regulations for fishermen to follow. These methods are used because they are effective in getting a large amount of fish in a short period of time saving fishermen time and effort. Some dangerous methods include over-fishing, blast fishing, bottom trawling, and cyanide fishing.

OVER-FISHING

Over-fishing is when fish are captured before they can reproduce, which can significantly reduce population sizes for the future. This disrupts not only the species that was harvested, but also the other organisms that depend on those species and potentially the whole ecosystem.

"Trawlers over fishing cod" by Asc1733 [CC BY SA4.0]

FACTS FROM THE WWF

- In just over 40 years there has been a decrease in recorded marine species by 39%.
- Around 93 million tons of fish were caught world-wide.
- Entire species, such as the Pacific Blue-fin tuna and swordfish, are highly endangered and are at an all-time low.

"Overfished US stocks 2010" by NOAA [CC BY-SA 3.0]

- Other marine species, such as whales, dolphins and turtles, are unintentionally killed as a result of over-fishing.

CAUSES

A major issue is the open access to the ocean and the absent regulations and monitoring of the water. Due to the increased number of fisheries, management has begun to slack. Current rules and regulations are not strict and do not mark a limit of intake. More importantly, there is little to no international fishing regulations. Even if the nations did come together to fight this issue, we still face illegal fishing. To reduce the intake of fish it should be mandatory to report intake and there must be a maximum limit that is determined by biologists and not companies. Also, we need stronger monitoring of the ocean, which may be hard due to the number of fisheries and the huge surface area the ocean covers.

IMPACTS

The WWF states that the populations that are mainly targeted are top predators in the ecosystem such as Billfish, Tuna, Salmon, and sharks. This is because of the economic and social demands of the fishing industry. Decreases in the top predator population can severely disrupt other marine populations. A prime example of this is increases in population sizes of smaller marine animals at the bottom of the food web that are fed on by top predators. This impacts other aspects of the marine ecosystem such as increases in algal overgrowth, which can be dangerous to coral reefs. Algae, although essential for the ecosystem, can have negative effects if there is a large abundance. Another issue that is closely related to over-fishing is by-catch. By-catch refers to non-target animals such as turtles or dolphins that are captured or killed in fishing nets. This threat causes the loss of billions of fish and other animals such as sea turtles and cetaceans.

"Sea Turtle Entangled in Ghost Net" by Doug Helton **[CC BY SA]**

Watch this Tedtalk on over fishing here and a video on by-catching here.

Over-fishing causes a cascade of effects in marine communities that can destroy habitats and result in the loss of biodiversity both in terms of overall abundance and species richness (Coleman, 2002). Not only does over-fishing destroy marine ecosystems, it also impacts food security for people. Humans that live in coastal communities rely largely on fish as a protein resource. Over-fishing decreases food security by threatening the long-term food supply, especially for individuals in developing countries.

Solutions:

There have been new movements to push fisheries to practice sustainable fishing. The WWF has helped develop and set environmental standards to help set a plan for sustainable fisheries. Approximately 15,000 seafood products hold to the standard of sustainable fishing, which is a great first step. There are other ways to help with over-fishing and one way is by influencing the market for fish. By reducing the need for fish products, less fish will be caught and hopefully it will allow some time for re-population.

By Drajay1976 [CC by 3.0]

BLAST FISHING

Blast fishing or dynamite fishing is a practice outlawed in most of the world, but is still used in southeast Asia. It involves using explosions to stun or kill large schools of fish for easy collection. The explosions often destroy underlying ecosystems from the strength of the blast. Around 70,000 fishermen still use this practice. Researchers believe that destructive fishing practices like blast fishing are one of the biggest threats to coral reef ecosystems. Coral reefs are less likely to grow in places of constant disturbance. The damage done to coral reefs has an immediate negative effect on the fish population in the area. From a single blast, it takes a coral reef about 5-10 years to recover. From constant blast fishing it leaves coral reefs unable to grow leaving an ocean of rubble. To reduce the use of this method, enforcement officials patrol the seas to try to catch and reprimand offenders.

Bottom trawling

Bottom trawling is a method that uses a large net that scrapes against the ocean floor to collect large groups of fish. Global catch from bottom trawling has been estimated at over 30 million tons per year, an amount larger than any other fishing method. The trawl doors disturb the sea bed, create a cloud of muddy water which hides the oncoming trawl net and generates a noise which attracts fish. The fish begin to swim in front of the net mouth. As the trawl continues along the seabed, fish begin to tire and slip backwards into the net. Finally, the fish become exhausted and drop back, into the "cod end" and are caught. The problem with bottom trawling is that it is un-selective in the fish it catches and severely damages marine ecosystems. Many creatures end up mistakenly caught and thrown overboard dead or dying, including endangered fish and vulnerable deep-sea corals that can live for hundreds of years or more.

Cyanide fishing

Cyanide fishing is a fishing technique used to gather fish for aquaria. In this process, a cyanide solution is used to stun fish for easier collection. This method can kill neighboring fish communities and severely harm coral reefs. Recent studies have shown that the combination of cyanide use and stress of post capture handling results in mortality of up to %75 of the organisms within less than 48 hours of capture. With such high mortality numbers, a greater number of fish must be caught in order to supplement post-catch death.

31 Chapter | PROTECTING THE REEFS ORGANIZATIONS

THE CURRENT CORAL SITUATION

With the current environmental crisis occurring on our planet, and the large amount of data we have to aid us in understanding the degree of the phenomenon, it is important now more than ever to utilize our knowledge to protect these ecosystems. Earth has a diverse array of natural habitats including Coral Reefs. It has been reported by the Global Reef Monitoring Network (GCRMN) that about 19% of the Earth's coral reefs are now dead, with multiple factors to blame including rising sea temperatures and ocean acidification.

Table coral of genus Acropora **by Yumi Yasutake via NOAA, 2008 under [CC BY 2.0].**

Coral reef ecosystems and the accompanying organisms that depend upon them are in danger of disappearing if we do not take action to protect these reefs. Coral reefs make up only 0.2% of our oceans, but they are home to over 25% of all marine fish species. The United Nations predicts that the Earth is on the brink of a massive extinction event, with some studies even suggesting 25% of the planet's species will be extinct by 2050. If we do not stop the effects of these factors we may lose a whole lot more than just the corals. The rapid decline of coral reefs will result in a significant number of social, economic and environmental tragedies around the world. We as humans must intervene knowing this current situation. In the US, the decline and danger from the loss of corals started to gain attention by the government in 1998. Since then, many reef protection organizations have emerged varying from government-funded to non-profits.

The USCRTF was established in 1998 by Bill Clinton who was President of the United States at that time. It's mission is aimed to protect and conserve coral reefs.

The USCRTF accomplishes this mission by mapping and monitoring the U.S. coral reefs, as well as helping to identify the problems causing the decline in reefs. Their goals as a task force is to find solutions to these problems, and to promote conservation and the sustainable use of coral reefs to the public. By working together with other organizations they try to find the best possible strategies to save the reefs. The USCRTF is responsible for many tasks which include but are not limited to:

- Implementation of Executive Orders
- Developing efforts to map and monitor coral reefs
- Research the cause and find solutions to these causes of reef decline
- Reduce cord reef degradation from pollution, and over fishing
- Implementing strategies to promote conservation and education to the public internationally

In 2000, the USCRTF adopted the National Action Plan to Conserve Coral Reefs *(National Action Plan).* This was the first blueprint for U.S domestic and international actions to address the issues of coral reef decline. This plan consists of conservation strategies to address the challenges that reefs are facing today. Not long after, the USCRTF developed the U.S. Coral Reef National Action Strategy *(National Action Strategy)* to further regulate the *National Action Plan. The National Action Strategy's* documents provide the framework for the priorities, strategies, and actions of the USCRTF and its members.

Similar to USCRTF there are many organizations out there trying to make a difference by saving coral reefs and the ecological communities that surround them. Below are four well known organizations trying similar tactics by identifying the problems and finding solutions for coral reef ecosystem protection.

Coral Reef Alliance (CORAL)

One organization working to protect our coral reefs is the Coral Reef Alliance (CORAL). They work with people from very different backgrounds: politicians, scientists, divers, and even fisherman. Their holistic conservation programs can be seen around the world, anywhere from Hawaii to Indonesia. After seeing the success of their programs many other communities and organizations have emulated their programs and efforts. Some of their work consists of creating healthy fisheries for reefs, assisting in ensuring clean water for reefs, protecting intact reef ecosystems, improving knowledge of the science of adaptation, among other things.

If you are interested in helping the Coral Reef Alliance there are many things that you can do. CORAL suggests a couple of things that can let you take action in your daily life such as traveling sustainably and working to educate others. If you want to take a more active role you can donate to their cause or even volunteer.

National Oceanic and Atmospheric Administration (NOAA)

NOAA by Wikimedia Commons **under [CC BY 2.0]**

A second organization working to protect the worlds coral reefs is the NOAA's Coral Reef Conservation Program. This organization was founded in the year 2000 by the Coral Reef Conservation Act and they have four main pillars of work: Increase resilience to climate change, reduce land-based sources of pollution, improve fisheries' sustainability, and to restore viable coral populations. In order to preserve the Coral Reefs they are working with many scientists involved in many different facets of NOAA. They believe that collaboration is the key to coral reef conservation. They partner with governments, academic institutions, and community groups to help save the coral reef environments. If you are interested in getting involved with the NOAA's coral reef conservation efforts there are a few ways you can do that. You can join the U.S. Coral Reef Task Force. If you are interested in heading your own efforts for Coral Reef conservation here is some information on how to get funded through the NOAA's grant programs. The NOAA also has job opportunities- here is a link to follow if you're interested. You can follow them on Twitter to see how they are getting the word out to the public.

Ocean Conservancy

Cleaning up trash **by** Wikimedia Commons**, 2013 under [**CC by 2.0**]**

Ocean Conservancy is another organization whose goal is to conserve every part of the ocean and marine life that surrounds it. This organization promotes health and diverse ocean ecosystems and aims to stop/limit the practices that threaten marine and human life. They focus on some of the main problems threatening marine life such as ocean acidification, trash, and helping set up sustainable fisheries to prevent over-fishing and the harmful effects of tourism.

National Fish and Wildlife Foundation (NFWF)

NFWF was created by Congress in 1984 and since then has become one of the world's largest conservation organizations. They work with both public and private sectors to help protect and restore the nation's fish, wildlife and habitats that surrounds them. They support conservation efforts in all 50 states and U.S. territories. NFWF has a primary focus on bringing together parties, getting results, and finding better solutions to help the future world. NFWF leads conservation financially with a funding budget up to $3.8 billion dollars.

Due to the major threats happening to coral reef ecosystems, NFWF has been responding to these problems by taking on multiple coral conservation initiatives in an aim to limit the threats, increase public awareness and find solutions to the problems of the decline of coral reefs both domestically and internationally.

To do this, NFWF works with many different partners to achieve these ideas for coral conservation. Some of these programs include programs that were discussed above such as managing the Coral Reef Programs with NOAA's Coral Program and USDA-NRCS. NFWF specializes in bringing all parties together (i.e. individuals, government agencies, nonprofit organizations and corporations). This promotes a solid foundation to protect and restore species, promote healthy oceans, improve wildlife habitat, help with sustainable fisheries, and water conservation. To date, NFWF has supported projects for coral reef conservation totaling over $34 million in about 39 countries giving them a global outreach.

REEF CHECK INTERNATIONAL

Reef Check International is an organization whose mission is to preserve the oceans and reefs emphasizing how critical they are to our survival. They work to not only protect tropical coral reefs but also California rocky reefs through education, research and conservation. Impressively, Reef Check has volunteer teams in more than 90 countries and territories.

The results of citizen science divers are used to improve the management of these critically important natural resources. Reef Check programs are notable for providing ecologically sound and economically sustainable solutions to save reefs, by creating partnerships among community volunteers, government agencies, businesses, universities and other nonprofits.

Dr. Karen Cangialosi of Keene State College is the Reef Check International Team Coordinator for the Turks and Caicos Islands. Read about her reef monitoring and youth education program here.

32 Chapter
ARTIFICIAL REEFS

UNIQUENESS OF CORAL REEFS AND THEIR STRUCTURE

Coral reefs are home to an abundant number of organisms and many different highly specialized species. Many of these species are endemic to these reefs, meaning they are only found here. Coral reefs also act as a nursery to some species that may not spend their whole life there. Reefs provide structure that promote safety and initiate the settlement of larvae of some species. Not only do they provide valuable services to the marine world, but also act as protection against storm swells and beach erosion for the terrestrial world (to learn more about corals visit here). However, they are in rapid decline as they face many challenges including climate change, ocean acidification and coral bleaching. So, what can be done about this decline? One possible solution is to construct artificial reefs.

WHAT ARE ARTIFICIAL REEFS?

Artificial reefs are man-made structures that are designed to imitate natural reefs. They imitate the physical structure that these reefs provide. They can be made from pretty much anything. Some are unintentional, such as oil rigs or boats that have sank, while others are constructed exactly for this purpose. Some are made from concrete or rock, others are made of metal or tires, but the most common object used to make artificial reefs are old ships. One example includes artificial reefs made from sunken New York city subway cars.

Structures are usually placed in open, featureless areas of ocean. These new structures act as a place for new coral polyps to settle, as well as larvae of many species. They can also act as a connection between different reef communities. Many marine organisms, including corals, have a planktonic larval stage. These larva have a set time and distance they can travel. If they don't settle before then, they will die. Artificial reefs can help connect different populations by serving as points for dispersing organisms to settle. These connections would be multigenerational in most cases as settlement means that individuals don't move from one reef to another, or in extreme cases, not at all. However, the offspring from one population could move to the next reef, and so on. Eventually, corals and sponges that are found in natural reefs take over and cover the man-made structures, and fish and invertebrates will be attracted to them like they are to natural reefs.

Subway cars being pushed into the water by unknown source under [SC DNR, CC 3.0]

Biscayne National Park by US National Park Services under [CC open access].

RESEARCH ON ARTIFICIAL REEFS

A study done by Arena, et al. (2007) that took place off the coast of Florida looked at the differences between natural and artificial (vessel) reefs. They found greater fish abundance and biomass on the vessel reefs, as well as greater species richness. There were also many economically important fish species found on artificial reefs. However, different assemblages of fish were associated with the different reef types. The vessel reefs housed a much greater proportion of planktivores compared to the natural reef. Planktivores on the younger vessel reefs are lower in the food chain, they represent resources yet to be unlocked and transferred to higher trophic levels. Over time, artificial reefs are expected to transition to be more similar to natural reefs as they age.

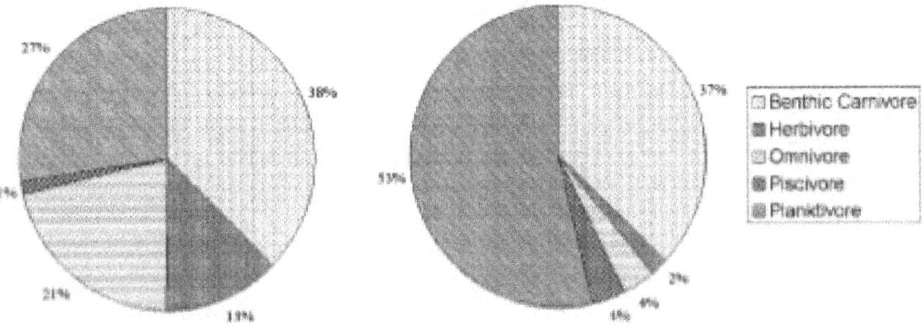

Figure 3 from Arena, Jordan and Spieler, 2007

Artificial reefs can provide a valuable solution to help with the threats that natural coral reefs face, but if things don't change and stressors continue to increase, then artificial reefs won't be sufficient.

There are many well known artificial reefs around the world primarily created from old machinery but there are also some that have been placed there initially for this purpose. "The Silent Evolution" by artist Jason de Caires Taylor, is a beautiful project where art sculptures were placed underwater to serve as artificial reefs. Visitors can enjoy the diverse underwater creatures but also take in the mystifying and magical art that is helping to protect Mexico's rich beauty and important ecology.

- Below is a video tour of "The Silent Evolution"
- Below is a video about Artificial reefs by Texas Parks and Wildlife:

It is worth mentioning that there are other types of reefs in the world besides coral reefs. Some, such as oyster reefs, can also be artificially made using simple materials such as oyster shells for spat (oyster larvae) to grow on and cement the loose materials into a solid reef structure. Many organizations such as The Elizabeth River Project and Chesapeake Bay Foundation work to restore oysters and reefs in the Chesapeake Bay and waterways of the US.

33 Chapter

BIOREMEDIATION TO CLEAN OUR OCEANS

WHAT IS BIOREMEDIATION?

Bioremediation is a possible solution for the growing problem of pollution and global warming. Bioremediation involves using living microbes to clean waste. These microbes can naturally occur in the area or scientists can extract them and place them into areas that need them. When scientists place cultured microbes into an area for purposes of bioremediation it is called bioaugmentation.

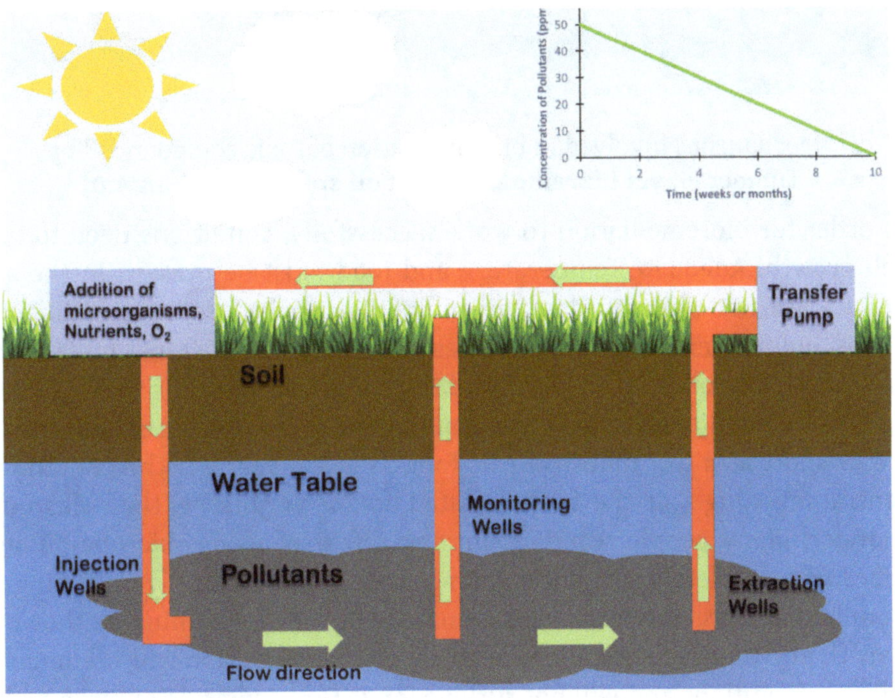

"In Situ Bioremediation" **by Jørgensen, K. S. via in situ bioremediation under** [CC BY-SA 4.0]

THE PROCESS OF BIOREMEDIATION

Microbes in the environment break down organic material. Some microorganisms can degrade contaminants by breaking down these materials into harmless substances. Bioremediation is about creating an ideal environment for microbes that degrade pollution by providing these organisms with fertilizer, oxygen, and other conditions that encourage their rapid growth. Microbes often break down pollutants into small amounts of water and harmless chemicals like carbon dioxide.

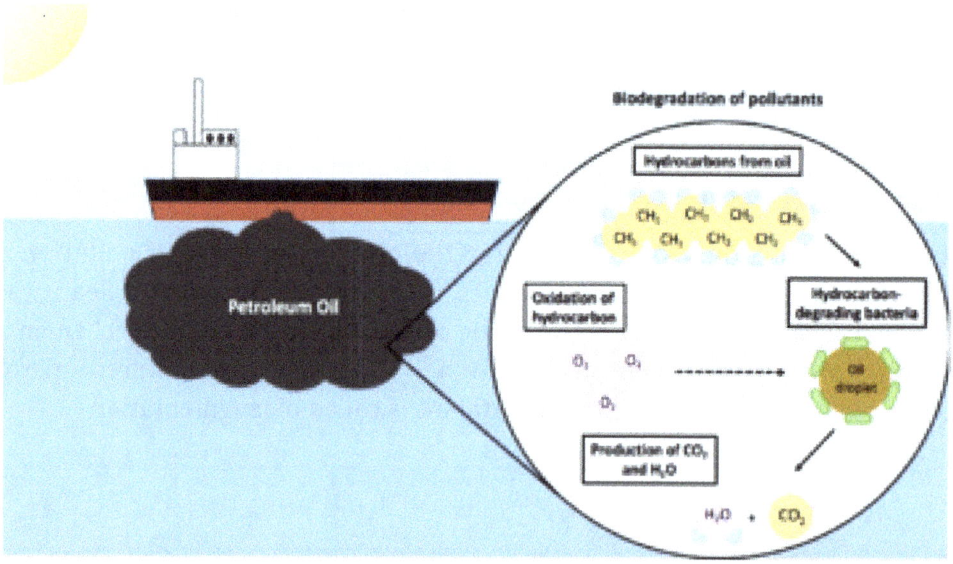

"Mechanisms involved in bioremediation of toxic compounds" **by** Timmer26**, via bioremediation of oil spills [CC BY-SA 4.0]**

In order for bioremediation to work successfully, conditions need to be ideal for cell growth. Adequate temperature and food must be present in the area. If conditions are not already ideal, amendments can be used. Amendments can consist of common materials like molasses or vegetable oil that increase the chance of survival for microbes by making conditions more suitable for their growth and enhancing their ability to break down contaminants. The process of bioremediation can be as short as a few months or as long as a few years. If the contamination concentration is high, or the contaminated area is very large, then clean up will take longer. Many people worry about the safety of using microbes. However, microbes are completely safe because they are already naturally occurring in the environment and are not harmful to humans. After the contaminants they needed for growth are depleted, the microbes from bioaugmentation will no longer have ideal living conditions and will not survive long. One downside of bioremediation includes creating noise pollution from the use of mixers and pumps that could be irritating to businesses and households.

Advantages of Bioremediation

Bioremediation is relatively cheap compared to other pollution cleanup methods. It is safe and can lead to cleaner water and soil in the area. Another advantage is that contaminated soil and groundwater are treated onsite without having to dig and transport materials elsewhere for treatment. One common example of a successful bioremediation process is composting. This video explains how bioremediation has been used to help clean up oil spills.

34 Chapter BIOCONTROL IN HAWAIIAN REEFS

WHAT IS BIOCONTROL?

Biocontrol or Biological Control is the introduction of a natural predator into an environment in order to control a pest species. It is an important way to aid in the management of terrestrial and aquatic habitats that have been invaded by non-native species. Using natural methods is a great way to help troubled ecosystems without many of the negative impacts associated with chemicals and other non-natural interventions.

Here is a video by the Hawaiian Conservation Alliance of the work completed in the bay of Oahu, HI for the urchin biocontrol project.

NATIVE SEA URCHIN BIOCONTROL EXAMPLE

Native sea urchins have been used as a form of biocontrol for algae overgrowth in Hawaiian coral reefs.

Two sea urchins in Hawaii by Opencage **under [CC BY-SA 2.5]**

A non-native species of algae was introduced to Hawaiian coral reefs in the 1970's that quickly became invasive and invaded Oahu's coral reefs. The algae grew so uncontrollably that it smothered out local flora and fauna and is still causing major issues for the Hawaii's reef areas today. Herbivorous fish are highly impacted by this overgrowth. People spent up to 8 hours a day pulling algae from the reefs in attempts to control its growth to little avail.

Finally researchers discovered the idea that native sea urchins could be used to control the algae. Urchins are great at controlling algal growth because they graze on the algae. After the introduction of 100,000 urchins to the reefs, an 85% decrease of algae cover was observed after just 2 years!

BIOCONTROL SUCCESSES AND FAILURES

Though there have been biocontrol failures highlighted in the media, such as the Cane Toad introduction to Australia, there have also been many great successes.

35 Chapter ECOTOURISM, RECREATION AND REEFS

Recreational activity is one of the joys in life that so many people share. Various types of recreation get people outside in all conditions from the 10o F New England winter weather to the hot beaches of a Caribbean island. Some tropical island activities include: boating, kayaking, scuba diving, snorkeling, sailing, wind surfing, and wake boarding- but they all have one thing in common, that is they are located in the ocean. Although some take place in deeper water, many of these activities are performed in shallow waters full of coral reef habitats. We know how fragile and important these reefs are and we also know how easily us humans have caused damage to them. Coral reef ecosystems are among the most biologically diverse and economically valuable ecosystems on Earth. Worldwide precious coral reefs attract millions of tourists annually and yield a significant economic benefit to those countries and regions where they are located. According to the National Oceanic and Atmospheric Administration (NOAA), recreation and tourism account for $9.6 billion of the total global net profit of coral reefs. This large amount of revenue generated is being threatened by the degradation of coral reefs.

As you can see there is a positive feedback loop occurring because of this situation. Many components of tourism, including recreational activities, are the cause of damage to the reefs, but ironically it has been shown that ecotourism is damaging as well.

Ecotourism is defined as: “Responsible travel to natural areas that conserves the environment and improves the well-being of local people.”

However, increased tourism to sensitive natural areas such as coral reefs, without appropriate planning and management, can threaten the integrity of ecosystems and local cultures. The increase of visitors to ecologically sensitive areas can lead to significant environmental degradation. Likewise, local communities and indigenous cultures can be harmed in numerous ways by an influx of foreign visitors and wealth. Mass tourism poses a threat to reefs and to the revenue generated from these ecosystems. Although branded under the word ecotourism many businesses and organizations are contributing to the increase in coral reef degradation. Once coral reefs are damaged, they are less able to support the many creatures that make their home on the reef and in turn lose value as a destination for tourists.

Little Venice quay flooded with tourists. Mykonos island. Cyclades, Agean Sea, Greece. Photo by Mstyslav Chernov. [CC BY-SA 3.0]

Most tourism in natural areas today is not ecotourism and is not, therefore, sustainable. Specifically, ecotourism possesses the following characteristics:

- Conscientious, low-impact visitor behavior
- Sensitivity towards, and appreciation of, local cultures and biodiversity
- Support for local conservation efforts
- Sustainable benefits to local communities
- Local participation in decision-making
- Educational components for both the traveler and local communities

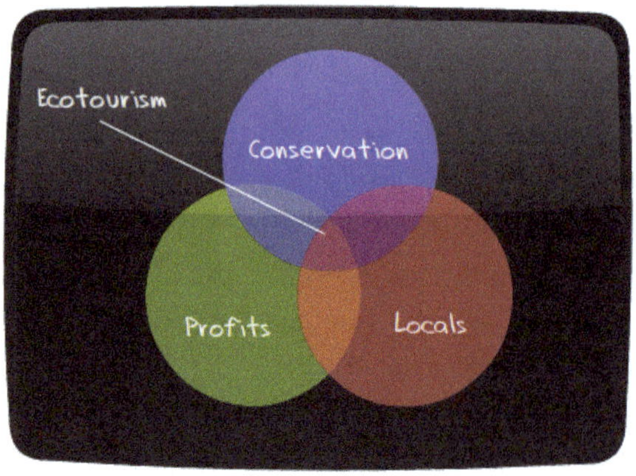

What is ecotourism? Photo by Ron Mader via Flikr. [CC BY-SA 2.0]

"Tourism will never be completely sustainable, as every industry has impacts. However, it's important to know if the revenue created from tourism is reinvested correctly in order to benefit the coral reefs and build a sustainable future. For ecotourism to be sustainable, companies must take responsibility and allocate revenue it generates from its eco-attractions into the protection of reefs instead of further investing in tourist structures and attractions that have negative impacts on the health of these ecosystems."

Recreational activities can harm coral reefs through:

- Breakage of coral colonies and tissue damage from direct contact such as walking, touching, kicking, standing, or gear contact
- Breakage or overturning of coral colonies and tissue damage from boat anchors
- Changes in marine life behavior from feeding or harassment by humans
- Water pollution
- Invasive species
- Trash and debris deposited in the marine environment

There are lessons to learn from the ecological destruction in Australia, Hawaii, Indonesia and other Pacific Islands where recreational activities are high in the bays of resort-filled areas and multiple-use marine parks.

In a study in Australia, activities such as diving, snorkeling, ski jets, and motor boats with surfing skis had high impacts on coral reef ecosystems. These activities can cause direct damage to the corals and increase pollution in the water. Surfing had less negative impact as it is superficial.

Some activities and their impacts are listed below.

[material below 1-4 is copied from How Does Tourism Affect Coral Reefs?]:

1. SCUBA DIVING AND SNORKELING

While most diving and snorkeling activities have little physical impact on coral reefs, physical damages to corals can and do occur when people stand on, walk on, kick, touch, trample, and when their equipment contacts corals. Coral colonies can be broken and coral tissues can be damaged when such activities occur. Divers and snorkelers can also kick up sediment that is damaging to coral reefs.

Divers. Photo by skeeze via Pixbay. [CC0 Public Domain]

Divers coming into contact with coral. Photo by Jjharvey8 via Wikimedia Commons. [GNU Free Documentation License]

Snorkeling on the Great Barrier Reef. Photo by Great Barrier Reef Encounter via Wikimedia Commons. [CC BY-SA 3.0]

2. BOATING AND ANCHORS

Boats grounding in coral reef habitat can damage corals, as can anchors. Anchors can cause a great deal of coral breakage and fragmentation, particularly from large boats like freighters and cruise ships. Heavy chains from large ships can break or dislodge corals. These damages to corals can last for many years.

Anchoring can also damage the habitats near reefs such as seagrasses that serve as nurseries and habitats for the juveniles of different coral reef organisms. Marinas may inappropriately dispose of oils and paint residues, polluting local waters, and additional pollution may occur during fueling.

Two brothers ship anchor. Photo by United States National Oceanic and Atmospheric Administration (NOAA)/Greg McFall via Wikimedia Commons. [Public Domain].

3. FISHING AND SEAFOOD CONSUMPTION

An abundance of tourist fishing and consumption of local fish stocks may lead to overexploitation and competition with local fishers. Inappropriate fishing techniques such as bottom trawling can cause physical damage to reefs.

CIMG2733 Fishing Net On Reef. Photo by Tim Sheerman-Chase via Flikr. [CC BY 2.0]

4. CRUISES AND TOUR BOATS

These vessels can cause physical damage to reefs through anchoring and grounding, as well as through the release of gray water and human waste into coral reef habitat. Chemicals added to paint used on boats and fishnets that are intended to discourage the growth of marine organisms can also cause pollution in coral reef waters.

Boats in the corals of Red Sea. Photo by kallerna *via* Wikimedia commons. [CC BY-SA 3.0]

Water Sailboat Sea Ocean Wind Sailing Yacht. Author unknown via Maxpixel. [CC0 1.0 public domain]

The variety of marine life and protected beaches supported by coral reefs provide beautiful sights for sightseers, sunbathers, snorkelers. Healthy reefs support local and global economies. Through the tourism industry and fisheries, coral reefs generate billions of dollars, and millions of jobs, in more than 100 countries around the world. Studies show that on average, countries with coral reef industries derive more than half of their gross national product from them. A good example can be found in Bonaire, a small Caribbean island. Bonaire earns about $23 million (USD) annually from coral reef activities, yet managing its marine park costs less than $1 million per year. A study conducted in 2002 estimated the value of coral reefs at $10 billion, with direct economic benefits of $360 million per year. For residents of coral reef areas who depend on income from tourism, reef destruction creates a significant loss of employment in the tourism, marine recreation, and sport fishing industries.

As we all know, coral reefs are undergoing major stress-related side effects because of human impacts. Through over-use, direct damage and ill-considered tourist operations, the World Wildlife Fund predicts that 24% of the world's reefs are under imminent risk of collapse through human pressures; and a further 26% are under a longer term threat of collapse. Another significant anthropogenic problem facing coral reefs is sedimentation. Sedimentation (losing soil from upland areas) is an extremely important cause of coral reef destruction. Coastal construction and shoreline development (back to the ecotourism concept) often result in heavy sediment loading. Watersheds cleared of their forests and other vegetation cover is vulnerable to erosion and flooding, resulting in increased levels of sediments reaching the reefs. Excessive sedimentation also exceeds the clearing capacity of some filter feeders and smothers the substrate. It reduces light penetration and can alter the vertical distribution of plants and animals on reefs. Sediments can also absorb and transport other pollutants.

When tourists accidently touch, pollute or break off parts of the reef, corals experience stress. The coral organisms try to fight off the intrusion, but this process often leads to coral bleaching—when corals react in a stressed way to expel the brightly colored algae that live in them this in turn, starves themselves and eventually become completely white. Once corals are bleached, they die and can no longer contribute to the biodiversity of the reef community. Since the disruption of one ocean system impacts all the others, sea grass and mangroves—shallow-water plant species vital to the health of the marine ecosystem—are also threatened by coral stress. Many of these events of accidental coral destruction are caused by recreational activities.

One study examined diver behavior at several important coral reef dive locations within the Philippines and also assessed how diver characteristics and dive operator compliance with an environmentally responsible diving program, known as the Green Fins approach, affected reef contacts. The role of dive supervision was assessed by recording dive guide interventions underwater, and how this was affected by dive group size. **Of the 100 recreational divers followed, 88 % made contact with the reef at least once per dive**. Divers from operators with high levels of compliance with the Green Fins program exhibited significantly lower reef contact rates than those from dive operators with low levels of compliance.

Although it's difficult for an individual to stop the entire coral reef dilemma it's easy to take small but powerful steps in the right direction.

Some of these steps include:

- Don't touch living coral and don't pick up wildlife for souvenirs, including shells, coral rubble and plants.
- Be conscious of what you bring with you, for example, reusable water bottles instead of plastic bottles and a backpack for your trash in case there isn't an area nearby to dispose of waste properly.
- Take the bus instead of a car, and if possible, do your research on the hotels or hostels where you stay.
- Try to stay at hotels that are environmentally friendly. Many coastal hotels dump their graywater—wastewater from laundry, cooking and household sinks—into the ocean, contributing to sedimentation and the contamination of coral reefs.

So, the message of this post is to be aware of corals and precious ecosystems when recreating! Also, try to vacation more sustainably by researching and traveling more eco-friendly. Some places that offer eco-tourism travel are green loons. Travel Tips for eco-traveling can be found below.

36 Chapter
HOW TECHNOLOGY IS SAVING THE WORLD'S CORAL REEFS

The Global Coral Reef Monitoring Network found that over 19% of the world's coral reefs are dead due to factors such as climate change and ocean warming causing an epidemic of coral bleaching across the world.

The decline in coral reefs causes major problems. Coral reefs make up only 0.2% of our ocean however they are home to over 25% of marine fish species and other organisms. They play many roles including protection of shorelines from major storms. The loss of coral reefs can disrupt many other ecosystems in and around the ocean.

Although the destruction of coral reefs have been caused by human activity, it is also humans that are saving these systems and finding solutions using technological advancements. Scientists from all over are exploring different technologies aimed at protecting coral reef ecosystems.

1. UNDERWATER ROBOTS THAT MIMIC OCEAN LIFE

Underwater robots may be a key in helping us understand coral reef systems. Underwater robots were recently developed by researchers at Scripps Institution of Oceanography at the University of California San Diego. This new tool can offer a new way to study ocean currents as well as the creatures that are in the ocean. Right now the goal of these underwater robots is to use them to help answer questions about the most abundant form of life in the ocean: plankton.

The researchers at Scripps have designed and built these underwater explorers to study small scale environmental processes that are taking place in the ocean. These robots include probes that are equipped with temperature and other sensors to measure ocean conditions by swimming up and down or by maintaining a constant depth. These small robots could potentially be deployed in the hundreds to thousands to capture a multi-dimensional view of the interactions between the physical ocean and marine life.

During a study using these robots, scientists wanted to test theories behind how plankton form dense patches under the ocean surface, which then later leads them to rise to the surface as red tides. These robots were the perfect way to mimic the underwater swimming behavior of plankton and examine the organisms' movements with the ocean currents.

This was the first time that a mechanism like this has ever been tested underwater. The biggest advancement is that these small robots are made inexpensively and are able to be tracked continuously underwater. What this means is that these robots could potentially be used as a small army and be deployed in a swarm. This swarm-sensing approach opens up a new world of ocean exploration. These small, low cost robots with cameras would allow for the photographic mapping of things such as corals. This technique could be huge in identifying the problem as well as seeing the effects of coral bleaching in almost real time.

2. ELECTRICAL BIOROCK STIMULATES CORAL GROWTH

To help preserve and restore coral, scientists are using an innovative technology called biorock. Biorock is a piece of technology that has a low-voltage direct current which is run through steel. This electricity can then interact with minerals in the seawater and cause solid limestone to grow on the structure. It uses the same principles as electrolysis, where the electric current causes a chemical reaction to occur that wouldn't normally happen. Coral fragments from other reefs can than be translocated to the biorock structure where they can grow due to the natural mineral crystals that were formed.

This type of technology is being used as conservation measures for coral reefs. This is because this type of technology can speed up the normal processes of coral growth. When a diver sees an injured coral they can move them to these structures so that they can continue to grow and heal. The coral then have 50% greater chance of survival then they had before. Biorocks have also helped fish and lobster populations, especially juveniles who shelter in the structures.

"Biorock Stimulation" **by** Ultra Kulture

3. 3D MAPPING AND BATHYMETRY TO MONITOR REEFS

Coral reef mapping and monitoring have helped scientists gather data on these habitats. With this technique scientists can collection biological, socioeconomic and climatic data needed to evaluate the conditions of the coral reefs and the surrounding area.

Monitoring operations include the National Coral Reef Monitoring Program (NCRMP), a new integrated and monitoring effort that will provide a precise picture of the U.S. coral reefs condition. To learn more about the NCRMP click here.

Mapping underwater coral reefs with 3D mapping allows volunteers and researchers to spend more time studying fish and invertebrates in water, while still getting the data needed from the reef. Cameras are used to take a multitude of underwater images and specialized software is used to analyze those images. More precise numeric data can be obtained with this technology. Before the use of technology such values could only be estimated by expert divers. Another advantage of this technology is its non-invasiveness, allowing the study of an environment in the office with less stressful field operations and coral extractions. The ease of repetitive surveys will make an amazing database for following coral reef changes over time, with the possibility to quantify all changes in near-real-time.

4. 3D PRINTED CORAL ENCOURAGES REEF RESTORATION

3D Printing Could Potentially Save our Coral Reefs: Check out this video showing the work being done to create artificial reefs

As coral reefs are declining at fast rates, scientists are struggling to find effective and innovative ways to save them. One of the new ways they are doing this is by 3D printing. 3D printing has become a common way to construct many things such as human organs. 3D printing of portions of reefs can replace lost pieces of coral. These "fake" reefs are thought to be less vulnerable to climate change and more resilient to changing environmental conditions. Scientists are using 3D printing technology that helps them to create fake reefs to mimic the texture and structure of the natural reefs in an effort for restoration. These experimental 3D printed reefs have already been implemented in the Mediterranean, the Caribbean, the Persian Gulf, and Australia. If they succeed, it will allow for new habitat for fish, but also baby coral polyps to attach themselves and multiply to grow into new reefs.

"3D Printed Coral Reefs" **by** Popularscience

Artificial reefs have also been used to provide habitat and act as an ecosystem by mimicking a coral reef. They have been made of sunken shipwrecks, plastic, concrete blocks, old tires, and old cars—all heaped onto the ocean floor in hopes that fish and other marine life will come to call them home. However, many of these reefs fail because they do not fit in. A 3D printed reef is a better option since it can recreate the small spaces needed to protect the species that live in the community, it can act as a passageway and door. It can also provide angles that cast light or shades in a certain directions enabling fish to avoid predation. Since this technology is so new there are still so many questions behind it like how well will they actually work? How sustainable are these structures? Can they withstand harsh storms?

As the years go on scientists will be able to determine the effectiveness of these structures. This technology is a step in the right direction to trying to find alternative ways to save coral reefs.

37 Chapter

INDIGENOUS PEOPLE AND CONSERVATION

THE PEOPLE AND THE LOCATION

Native peoples living near coral reefs have been displaced for centuries for many reasons some including tourism, agriculture, and conquering of land. Today, nearly 40% of the world's 6,000-8,000 indigenous groups encompass coastal ocean and island regions within their homelands, territories, and nations.

Below is a list of some native populations that live/lived near reefs around the world

- The Ciboney and Carib peoples from mainland America inhabited some land the West Indies.
- Giraavaru people were the ancient owners and rulers of the Maldives.
- The Taíno were an Arawak people who were the indigenous people of the Caribbean and Florida.
- Aboriginal people and Torres Strait Islander people belong to Australia and are connected with the Great Barrier Reef.
- Sentinelese people an indigenous people who inhabit North Sentinel Island in the Bay of Bengal in India. They are considered one of the world's last uncontacted peoples.
- The many tribes of Papuans living on Papua New Guinea.
- Malagasy people of Madagascar make up many different groups.

Below is a map of coral building reefs around the world. As seen they are located along the tropics where high densities of human populations live.

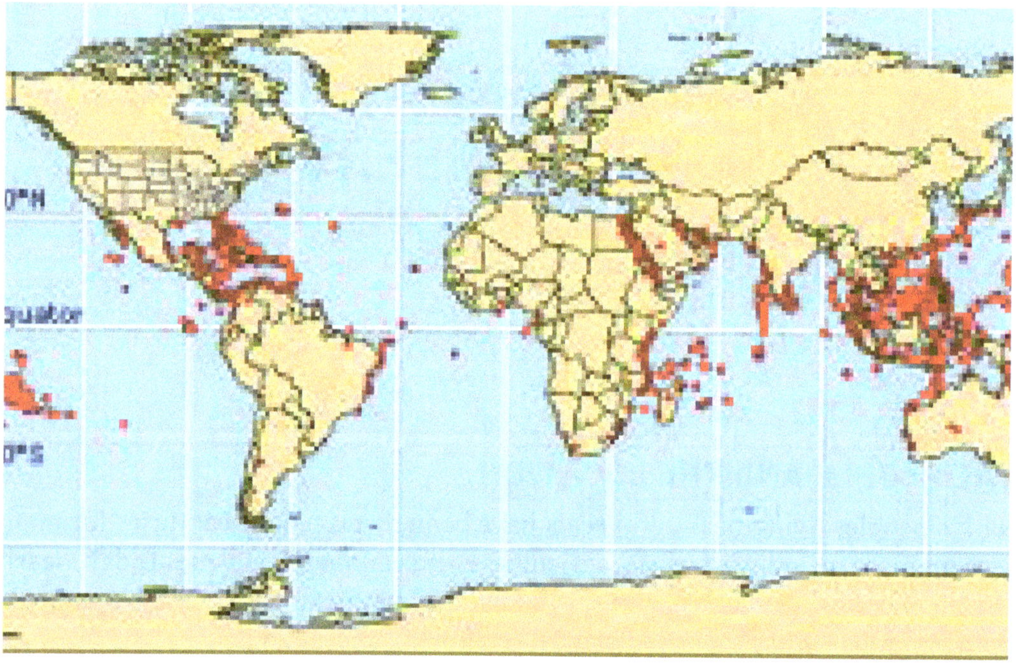

"Reef building corals" **by** National Oceanic and Atmospheric Administration **licensed under** [public domain]

CLIMATE CHANGE THREATS

Many inescapable threats besides displacement affect indigenous cultures today. One heavily discussed is Climate Change. Climate change is impacting many indigenous communities by endangering sacred and traditional living sites, cultural practices, local forests and ecosystems, traditional foods and water quality. In response to this crisis, scientists in the US are attempting to work with coastal communities to study the impacts of climate change on the health and vitality of the social, economic and natural systems of these communities.

As climate change affects coral reefs intensely, these impacts influence the people that live near and depend on these ecosystems. Two studies in Hawai`i are exploring climate change impacts on coral reefs and in particular the flooding of coastal communities from rising sea water levels. It is agreed widely by scientists that climate change poses the greatest long-term threat to coral reefs and among other impacts, climate change is expected to result in more frequent severe tropical storms and severe coral bleaching events. Coral bleaching is caused by environmental stress from warm water which causes corals to expel their symbiotic algae from their tissues, turn white and eventually the afflicted coral may die. Threats from increasing storms, rising sea level and direct effects of ocean warming will continue to have devastating impacts on coastal ecosystems and their human communities.

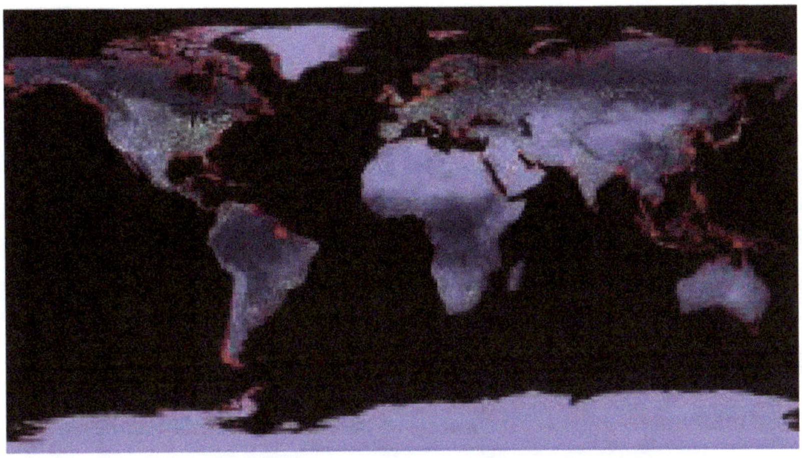

"6M Sea Level Rise" **by NASA under** [public domain]

INDIGENOUS SOLUTIONS

Scientists, along with members of Washington State's Swinomish Indian Tribal Community, are addressing climate change threats to the Tribe's land, wildlife, culture and community health. As sea levels rise, coastal erosion intensifies which lead to many negative side affects on land and in the ocean. One important part of preparing for climate change is identifying a full range of potential impacts from declining natural resources such as fish, to damaged infrastructure such as roads and buildings, to compromised community health. The Swinomish fishing tribe in Seattle WA are taking action to build community resilience by implementing the Swinomish Climate Change Initiative. With funding from the Northwest Climate Science Center and the North Pacific Landscape Conservation Cooperative, the Swinomish have conducted a pioneering study to combine assessments of ecological health with newly developed community health indicators to identify priority adaptation tactics. For more information watch a webinar about the project here. Other funding has been granted towards these areas such as from the Interior Department which recently claimed to make $8 million available to fund projects that promote tribal climate change adaptation and ocean and coastal management planning through its Tribal Climate Resilience Program.

On the island of Maui scientists are developing a tool to help coral reef managers make science-based decisions; this tool can map, assess, value and simulate changes in the coral reef ecosystems under different climate change scenarios. Hawai`i's coral reefs provide seafood, areas for recreation and tourism, coastal protection and support the traditional lifestyles and values of the state's native cultures. The research will show potential climate change impacts and help land and coastal managers make informed decisions to create resilient coral reefs and coastal communities.

Divers from the Commonwealth of the Northern Marianas Islands reef resiliency team conduct assessments of reef resilience in the Marianas archipelago by USGS licensed under [public domain].

THE BATTLE OF GOOD VS GOOD

Traditionally native peoples were displaced due to other nations conquering the land and many now assume it is all over. Unfortunately, this problem still occurs today with a strange and complicated twist. Many indigenous people are being pushed out of their homes in the name of conservation because governments can acquire money for conservation projects like setting aside land. When the economic priority is to generate revenue from conservation, humans get pushed away from the protected areas so they can become "protected for nature". Most of the world's 6,000 national parks and 100,000 protected places have been created by the removal of tribal peoples. Hundreds more parks are being created every year as countries commit to meeting the UN's goal to protect 17% of land by 2020. And the human toll is rising accordingly.

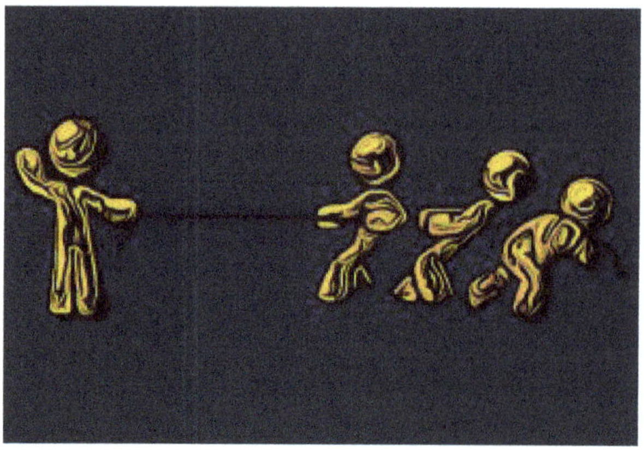

Resist tug of war: **by author unknown via** Pixabay, **July 9, 2019.** [Simplified Pixabay License].

Popular Books Written on this Topic

Misc. Resources

- MinorityRights.org provides information on indigenous peoples in all countries and regions on the world.
- Reefbase.org is a coral reef database full of information open to all

38 Chapter | VOLUNTEER TO BE VICTORIOUS

WHAT IS VOLUNTEERING?

There are countless people and organizations working around the world at this moment to help protect coral reef ecosystems. Many of these people are unpaid volunteers and realize the importance of coral reef conservation and its effects on everyone around the world. Volunteering is an amazing form of work that anyone can do anywhere in the world, within any area. It is simply helping out or working without compensation other than self-satisfaction. According to Nationalservice.org the 2018 Volunteering in America report found that 77.34 million adults (30.3 percent) volunteered through an organization last year. Altogether, Americans volunteered nearly 6.9 billion hours, worth an estimated $167 billion in economic value, based on the Independent Sector's estimate of the average value of a volunteer hour for 2017. Millions more are supporting friends and family (43.1 percent) and doing favors for their neighbors (51.4 percent), suggesting that many are engaged in acts of "informal volunteering". There are so many different environmental organizations one can volunteer with and below is a list of just a few that involve coral reef projects.

Organizations and information from their websites:

Mission: Our mission is to change the face of volunteer travel. Established in 2007, we have grown to become the world's leading volunteer travel company, working in over 40 destinations around the world and placing thousands of volunteers abroad every year.

We believe in a future where any traveler, anywhere in the world is empowered to make a meaningful difference in the community they are visiting, and we take pride in making this happen.

We're focused on providing affordable volunteer travel experiences that are responsible, safe and high quality. Our programs heighten global awareness and cultural understanding through the skills and expertise taken by volunteers to their host communities, and through the experiences and lessons that volunteers take back to their own countries and cultures.

Price: Fees from $1010 for 1 week

Time Commitment: 1-4 weeks

Tasks you would do!:

Volunteers on the Marine Conservation project in Australia have the opportunity to join a variety of important conservation efforts focused on the protection of Great Barrier Reef ecosystem.

When volunteering in Australia, you will work in collaboration with a number of oceanographic organizations to gather vital raw data and support the protection of the Great Barrier Reef through a range of initiatives, including:

Reef Monitoring – This portion of the Marine Conservation project involves snorkeling within an assigned area to collect data on the species living in the Great Barrier Reef. You do not require any previous reef surveying experience to participate, as you will be trained in the methodology of in-water surveying during your program orientation. A full-length lycra suit will be supplied and volunteers are just required to bring their own snorkel and waterproof watch.

Name: GVI

Mission: 20 years later, GVI has engaged over 35,000 participants, set up 600 community partnerships, and currently runs 21 programs in 13 countries worldwide. Richard and Ben's vision to not only facilitate global citizenship and leadership skills in young adults, but to allow them to have a truly positive impact on local communities and environment is very much alive.

TIME COMMITMENT: 4, 6, 8, 12 WEEKS

Tasks you would do!: Travel to the crystal clear waters of the Indian Ocean as a member of an expedition and work on critical marine conservation projects amongst the beautiful islands of the Seychelles.

You will contribute towards various conservation-related surveys aimed at providing data to the local government on coral reef research, fish, and invertebrate surveys and assist with the development of an environmental education and awareness program as well as marine plastic pollution cleanups and surveys.

You will spend the majority of your time on this expedition scuba diving and as such you need to be qualified to at least PADI Open Water, or equivalent. For non-divers wishing to attend, we can recommend local dive centers that will help you qualify before your intended start date.

Name: Earth Watch

Mission: Earthwatch engages people worldwide in scientific field research and education to promote the understanding and action necessary for a sustainable environment.

Price: $1550

Time Commitment: 5+ days

Tasks you would do!: On this expedition, participants can get involved through scuba or snorkel activities. You will assist researchers in making baseline measurements of environmental conditions, actively removing algae, deploying coral recruitment (settlement) tiles, and assessing fish and invertebrate diversity and abundance. By joining this expedition, you'll be at the forefront of active reef restoration science. You will assist researchers in experiments that aim to develop best practice methods for removing this macroalgae and allowing coral to regrow. You will be directly involved in filling in the gaps that will enable reef managers to make evidence-based decisions about active interventions that support the recovery of the Great Barrier Reef, and reefs all over the world.

Name:OPERATION WALLACEA

Mission: Operation Wallacea is a network of academics from European and North American universities, who design and implement biodiversity and conservation management research expeditions.

Research is supported by students who join the programme, to strengthen their CV or resume or collect data for a dissertation or thesis. Academics benefit from funding for high quality fieldwork enabling them to publish papers in peer reviewed journals. This model enables the collection of large temporal and spatial datasets used for assessing the effectiveness of conservation management interventions.

Price: $5,925.00 ($2,370.00)

Time Commitment: 8 weeks

Tasks you would do!: IVHQ's affordable Sea Turtle and Marine Conservation volunteer projects offer international volunteers with the opportunity to provide vital support to ocean conservation organizations around the world that are focused on the protection of fragile ecosystems and threatened marine species.

BENEFITS OF VOLUNTEERING

With busy lives, it can be hard to find time to volunteer. However, the benefits of volunteering can be enormous. Volunteering offers vital help to people in need, worthwhile causes, and the community, but the benefits can be even greater for you, the volunteer. Giving to others can also help protect your mental and physical health. It can reduce stress, combat depression, keep you mentally stimulated, and provide a sense of purpose. While it is true that the more you volunteer, the more benefits you'll experience, volunteering does not have to involve a long-term commitment or take a huge amount of time out of your busy day. Giving in even simple ways can help those in need and improve your health and happiness.

Just a few benefits of volunteering are listed below

- Connects you to others
- Good for your mind and body
- Advance your career
- Brings fun and fulfillment to your life
- Doing things you would not do otherwise

Chapter 39 KEENE STATE COLLEGE OUTREACH

TROPICAL MARINE BIOLOGY CLASS

The Tropical Marine Biology Class taught by Dr. Karen Cangialosi at Keene State College, allows undergraduate students the opportunity to investigate topics such as tropical marine ecology, biodiversity, coral reef biology, conservation, and environmental issues. The course incorporates a rich field trip experience to the Turks and Caicos Islands where students get direct hands-on experience diving and snorkeling to observe marine life. Students also learn the methods of reef monitoring and learn much about the local culture of the islands. Tropical Marine Biology students continue to play an important role as instructors in the September Reef Education Program.

BEYOND THE CLASSROOM BY JAIME MARSH

Picture this: you sign up for a class, on the first day you take a seat in a generic, oversized lecture hall alongside 20 other classmates, you open your textbook to the first chapter, the professor walks in and shortly begins lecture. With the exception of a few monotonous exams, this is how the next 15 weeks of the semester will be. This is how the majority of classrooms have operated throughout our middle school, high school, and college years. For many of us, we do not know any differently, however a new wave of an open education based curriculum is slowly approaching that is very quickly changing the way we learn.

For me, it started in a Tropical Marine Biology course I enrolled in, in my junior year at Keene State College. We began the first day of class in the most non-traditional way possible, by being asked one simple question: 'what do you all want to do this semester?' While we still had some traditional components to the course like lectures, it rapidly evolved into something much larger- from something we learned, to something we could apply, to something we created, to something that I love. While the series of identification quizzes at the start of the semester provided me a solid base to apply my knowledge of certain species and marine ecosystems, they did not prepare me for the depth and wealth of my actual experiences on the May 2017 trip. For many reasons, this trip to the Turks and Caicos Islands forever changed me.

First and foremost, it provided me the opportunity to return to TCI in September 2018 to participate in the Coral Reef Monitoring and Youth Education Program under the direction of Dr. Karen Cangialosi and Dr. Scott Strong (read more about this below).

Once there, my classmate Alana and I assisted in the collection of data for the Reef Check International database, as well as helped run a small after-school snorkeling program for a few of the local high school girls. With my first trip to TCI under my belt, I was confident and excited to not only share my knowledge of the coral reef system, but to also share my passion for the ocean and all of its' wonders. However, to say it was a humbling experience would be an understatement. Dr. Cangialosi had previously stated that many of the Turks Islanders do not know how to swim, and in fact, many have never even been in the ocean because they are scared of it or have been discouraged from swimming. When we asked the girls if any of them knew how to swim, most of them timidly answered 'yes,' however once in the water, their lack of experience was evident. For the first few days we focused largely on teaching the girls not only how to swim and float, but also teaching them the basics of the coral reef system and species identification. The majority of the time they clung to us; my hand had never been so numb. Several times I was pulled underwater, or hit in the face by flailing arms and flippers, all while treading water for over an hour. Over the course of a week we watched them become more comfortable, not just with the ocean and the water, but also with themselves. They slowly let go of our hands, and before you know it heads were popping up out of the water yelling to us in pure excitement that they had just seen a Queen Parrotfish, a Spotted Eagleray, or their favorite-a Sea Turtle. It was incredibly rewarding to see these girls begin to excel so quickly at something they were at first so afraid of.

I felt that our time with this group of girls went beyond what we originally set out to do. The goal was to teach them how to snorkel and give them information regarding the coral reefs; but within this short period of time, the moments that mattered the most were the small victories that each girl accomplished. One girl floated for the first time all by herself, another swam for 50 meters without stopping, and all of them were identifying the different species that comprise the coral reef system. We also listened to their hopes and dreams, their plans after high school, and encouraged them to go to university. I would like to think that we instilled them with confidence to go forth in life and follow any dreams big or small. And while I would love for any of them to continue to snorkel and identify the different species, and even share their knowledge with others, I do not believe that this was the biggest take away. They learned to not only trust themselves, but to trust a complete stranger, to be independent but to lean on others when you need to, and most importantly I think they learned how to believe in themselves when faced with a challenge.

Girls in Keene State College-TCI Reef Education Program, photo by Scott Strong, copyright.

That being said, it would be naïve of me to say that they did not have a similar impact on me because while this experience was humbling, it was also rewarding. The number one lesson I learned was that there are just some things that you can't learn in a classroom. Yes, in the classroom I learned how to identify at least a hundred different types of marine life, but now knowing that I can actually apply it outside in the real world puts many things into perspective. To begin with, a classroom can't teach you that everyone comes from a different background, and it is likely different than what we see on the surface or what we imagine. A classroom can't teach you that it's important to find the joy and happiness in all moments, no matter how big, small, sad, or happy they may be. A classroom can't teach you what you love, just so you can do it, you have to experience it-whether it is medicine, marine biology, teaching, or all of the above. A classroom can't teach you that it is alright to go down a path to find out it's not where you want to be-sometimes you just have to travel down it. Some lessons are meant to be learned outside of the classroom, even if it happens to be in the middle of the ocean, 60 meters underwater. In the famous words of Jacques Cousteau, "The sea, once it casts its spell, holds one in its net of wonder forever."

CORAL REEF MONITORING, YOUTH EDUCATION IN THE TURKS AND CAICOS ISLANDS

Photo by Scott Strong (copyright, use with permission)

In 2008, Dr. Karen Cangialosi and Dr. Scott Strong, established a coral reef monitoring program off the shores of Providenciales, in the Turks and Caicos Islands (TCI). On an annual basis habitat quality, measures of the physical environment, and invertebrate and fish diversity along a linear 100M transect on a small patch reef in Grace Bay are collected for data using the protocols of Reefcheck International. Each year the data is submitted to the Reefcheck International database, an Open Dataset available to the public. The data that is collected is also submitted to and used locally by the TCI Department of Environmental and Maritime Affairs. Over the years, they have established strong relationships with many individuals on Provo including business owners/operators, government officials, teachers, and school administrators. Dr. Karen Cangialosi serves as the Reef Check Team Coordinator for the Turks and Caicos Islands.

Integral to this project, they also run a small-scale Reef Education Program (REP) with resident students affiliated with the Gartland Youth Centre on Providenciales (Provo). This program is designed to help youth develop an understanding of the marine environment and emphasizes that sustainable practices can ensure not just the health of TCI's reefs, but lead to healthier and more equitable living for its people. REP participants (14 to 17-yr-olds) are recruited primarily from Clement Howell High School, which has approximately 350 students including local Turks Islanders (Belongers), and expatriates primarily from Haiti, Jamaica, and the Dominican Republic. Participants learn snorkeling, marine life identification, behavior, diversity, and ecology of coral reefs, while also focusing on issues central to reef conservation. This program has been very successful in contributing to increased environmental awareness, internships, jobs, university admission and other opportunities for our Turks Islander REP participants.

Chapter 40 PHYLUM PORIFERA

ORGANISMS

Porifera

Sponges are unique creatures. They are in the Phylum Porifera and there are about 5,000 different known species. They are one of the simplest forms of multi-cellular animals and come in a variety of different colors, shapes, and sizes. Sponges lack organs and a nervous system. They are sessile organisms, attached to reef surfaces via a holdfast. They get their food through filter feeding when the ocean currents go through their pores. There are 4 different classes of sponges; Calcarea (calcareous- has spicules*), Hexactinellida (horn sponges), Demospongiae (coralline), and Sclerospongiae (glass sponges). Each class of sponge is composed of different organic materials. If you want to learn more about sponges and their structure, this is a really informative article about them!

"Venus Flower Basket Glass Sponges" by NOAA [CC by 2.0]

Spicules are "sharp spikes made of calcium carbonate located in the mesohyl. Spicules form the skeleton of many sponges". Spicules can be a form of defense for the sponges to deter or hurt predators.

Sponges are some of the oldest organisms on earth having been around up to 500 million years! Each individual cell in a sponge can transform to do any function within the sponge. This is a unique and helpful trait to have. An article from Oceana.org even says that "a sponge destroyed in a blender can reform itself as the cells swim back together and take on the form and job needed for recovery".

Most sponges are also hermaphrodites meaning that each adult can act as either male or female in the reproductive process. Some sponges can also go through asexual reproduction to produce clones when a piece of sponge breaks off and grows in another location!

Sponge fibers help filter water through the organism. This is important because they filter feed on plankton and bacteria while attached to the ocean floor

There are many different types of rope sponge. Row Pore Rope Sponge, Scattered Pore Rope Sponge, Thin Rope Sponge, and so on. They are a tropical sponge that come in a variety of colors, sizes, and depth ranges. This Marine Species Identification portal has some amazing information on marine species.

"A red rope sponge (Amphimedon compressa) in Hawa'ii" by Laurie Minor-Penland [CC BY 2.0]

Tube sponges are another type of tropical sponge. They have a tube shape and also come in a variety of colors, shapes, and sizes. Below is the Yellow Tube Sponge, also known as Aplysina fistularis. The Yellow Tube Sponge has a large opening at the top of it and is attached at the bottom. If they are exposed to the atmosphere, they will turn purple and black and die.

"Aplysina fistularis– Yellow Tube Sponge" by Nick Hobgood [CC BY 3.0]

The Barrel Sponge is another common type of tropical sponge. They are called Barrel Sponges because of the way they look. They are bowl shaped and have very large openings (osculum) at the top of them. They grow to be at least 6 feet wide, making them one of the largest sponge species in their habitats. They are homes for a variety of marine life such as shrimps, crabs, gobies, and cardinalfishes.

"Xestospongia muta, the Barrel Sponge" by NOAA [CC by 2.0]

Fun fact: Giant tube sponges can live up to 2000 years old!!

Sponges are home to an array of microorganisms. Green algae, cyanobacteria, archaea, cryptophytes, red algae, dinoflagellates, heterotrophic bacteria, and diatoms are known to inhabit a range of sponge species. Most microorganisms that regularly inhabit sponges live in symbiosis with the sponges, however parasitic and pathogenic organisms can also affect them. Some microorganisms give corals their vibrant colors. Multiple microorganisms may live in the same sponge at the same time as well. The symbiotic relationship can be either commensalistic or mutualistic, and the types of microorganisms found in and on sponges can have an effect on the sponges growth and development, as well as its color.

41 Chapter

PHYLUM CNIDARIA

Cnidarians are organisms found exclusively in aquatic habitats. Of the over 10,000 aquatic Cnidarian species discovered, most inhabit marine environments. Cnidarians exhibit two major body forms, polyp or medusae. Polyps consist of a body stalk with a tubular shape with a single opening and multiple tentacles that surround this opening which serves as a mouth and anus. Polyps point upward for filter-feeding. The tentacles are covered with cnidocytes (stinging cells). Cnidocytes are the most important characteristic that distinguishes organisms as belonging to the Cnidarian phylum.Polyps are sessile and remain attached to a substrate. The medusa form is most notably observed in the umbrella like form of jellies. Similar to polyps, they consist of a body with a single opening surrounded by tentacles, but the gelatinous layer is much thicker and the mouth is usually oriented towards the substrate when swimming. Many Cnidarians have the ability to switch between these two morphs over the course of their life.

CLASS HYDROZOA

Hydrozoans can be found in both the medusae and polyp form in equal abundance. Compared to other cnidarians hydromedusea have a thinner and more delicate mesoglae (gelatinous) layer. In the polyp phase, most hydrozoans live colonially and often have polyps that bud from other polyps. They have a shared gastrovascular (GV) cavity and are genetically identical. Since the GV cavity is shared if one polyp is actively carrying out its function, such as feeding, the product of digestion (acquisition of nutrients) can be distributed among the other polyps. Each polyp can have a different function in the colony. Although they are all genetically identical different genes are expressed to vary the function of the individuals. Individual polyps are known as zooids. Gastrozooids, gonozooids, and dactylozooids respectively function for digestion, reproduction and defense1.

Hydras are hydrozoans that are unique in several respects. Unlike almost all other hyrdrozoans, they are solitary, live in fresh water and lack a medusa phase. Like many Cnidarians, Hydras can contain endosymbiotic algae or Zooxanthellae.

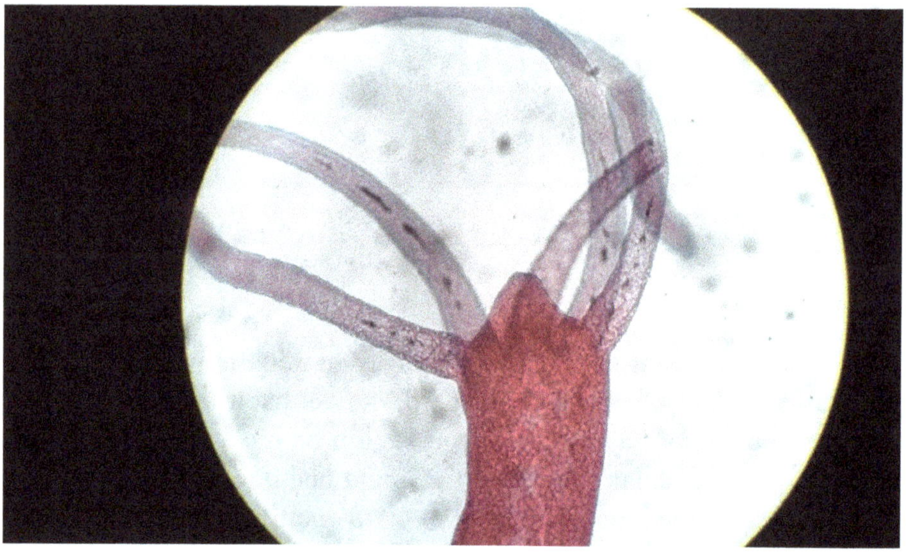

"Hydrozoan Polyp under Compound Microscope, 2019" by Jason Charbonneau [CC by 4.0]

Some benefits of being an individual within a colony are protection, resource sharing, and even the passing of signal impulse. If one individual is stimulated by a threat, the signal will be sent throughout the colony, initiating a retraction across the whole colony.

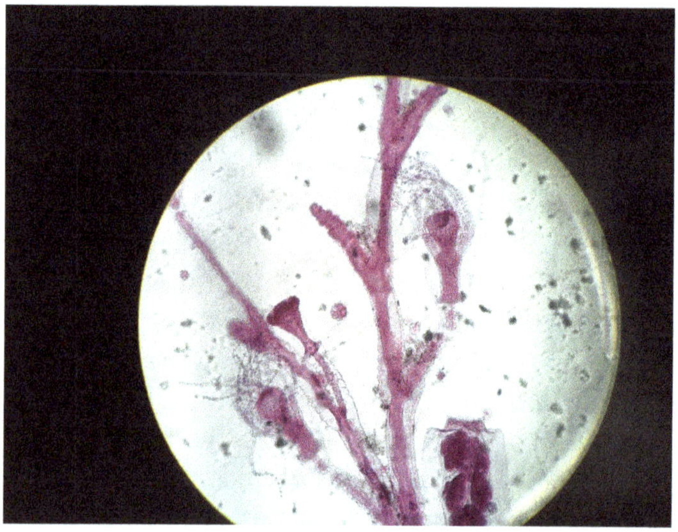

"Obelia (colonial Hydrozoan) under compound microscope, 2019" by Jason Charbonneau [CC by 4.0]

Class Anthozoa

Anthozoans only occur in the polyp body form. Their polyps are more complex than the other Cnidarian classes, and the colonies can grow to enormous sizes. Anthozoans include the hard corals, gorgonians, soft corals and sea anemones. Read more about corals in chapter 3.

When analyzing the "skeleton" of corals, there are distinct factors that you should pay special attention to. When they grow, coral polyps deposit CaCO3 which is constantly accumulated around the living organisms throughout the course of their life-cycles. The live polyps are housed within the calcareous cups that they secrete, these serve a skeletal function and for protection. Analyzing the 'skeletons' left behind after the death of these organisms, you can see how they lived and developed over time. All of the perforations that might be observed in the dead skeleton were once home to a live polyp that made up the actual live being of a coral.

The photo below depicts a living hard coral polyp with its cnidocyte-covered tentacles protruding from the calcareous cup.

"Coral Polyp under dissecting Microscope, 2019" (Cnidocytes are visible as bumps on the tentacles) by Jason Charbonneau [CC by 4.0]

CLASS SCYPHOZOA

Scyphozoans are found predominately in the medusae form, but also have polyp phases in their life cycles. The spectacular variety of jellies are in this class. Check out the Monterey Bay Aquarium's fantastic Live Jelly Cam!

Sea Nettle (Chrysaora fuscescens) Jellyfish **in captivity in the** Monterey Bay Aquarium **[CC BY SA 2.5]**

REFERENCE

Brusca, R.C., & Brusca, G.J.(2003). Invertebrates. Sunderland, MA: Sinauer Associates, Inc.

42 Chapter PHYLUM ANNELIDA

ANNELIDS

The Phylum Annelida is defined by its morphology as segmented worms. They express closed circulatory systems in the same fundamental manner that *Homo sapiens* do. Annelids are categorized taxonomically into many classes. A few of these classes are: Class Polychaeta, Class Clitellate and Class Echiuran. Common organisms in the Class Clitellate include Oligochaeta (earthworms) and Branchiobdellida ad Hirudinea (leeches). Ecologically, Annelids are subdivided into two classifications. Errent polychaetes express a multitude of parapodia appendages, defined heads complete with proboscis and are able to extend their bodies in order to have a greater physical surface area for gas exchange. The latter group being Sessile Tube Dwelling Annelids. These organisms contain the segmented part of their bodies within an external tube. The portion of their being that is extended is a feeding appendage that filters through the water. A great example of a Sessile Tube Dwelling Annelid is the Christmas-tree worm. Each Christmas-tree worm has two feeding appendages to filter through the water. They also have the ability to retract quickly when the worm feels threatened. Some are even photosensitive, which means when natural light is obscured by a potential predator, they automatically retreat. Sessile annelids are also inclusive of medusa worms and deep-sea thermal vents.

"Christmas Tree Worm on Brain Coral" by Alana Olendorf [CC by 4.0]

Split-crown feather dusters also live in reefs like the Christmas-tree worm. They have two semi-circle crowns of radioles which form a crown with a split down the center. These worm don't just live on the reef, they live in tubes that the worm builds from particles of sand and a "glue" that is excreted from the worms body. The crown of radioles act as gills and that capture plankton and other microscopic organisms that the worm can feed from. When threatened, feather dusters will quickly retreat back into their tubes.

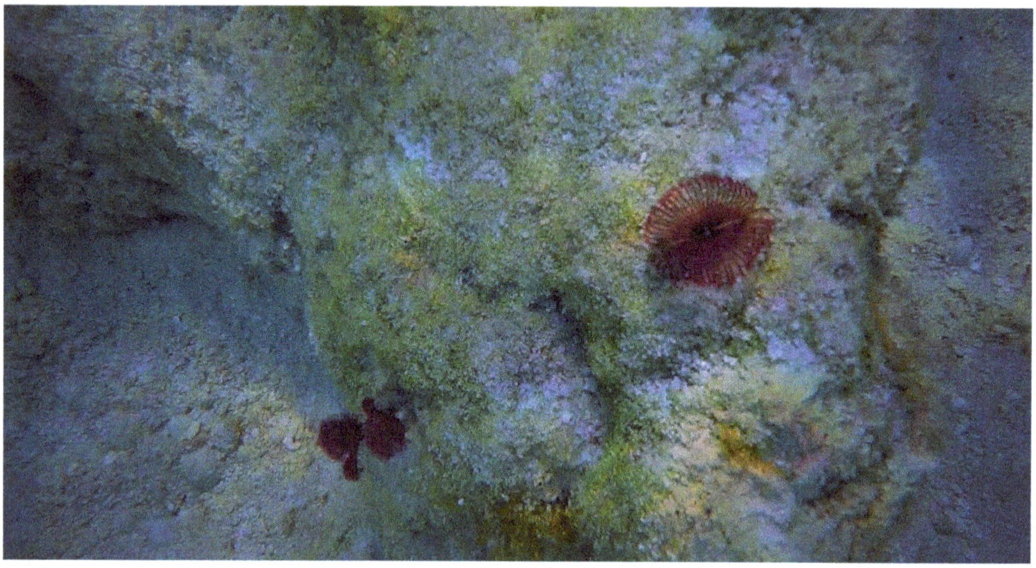

"Split-crown Feather Duster" by Alana Olendorf [CC by 4.0]

In the Turks and Caicos and other locations where they are found, there is an incredible phenomenon that takes place with the *Odontosyllis enopla*, otherwise known as the Bermuda Fireworm. This is a small, polychaete worm found in the western Atlantic Ocean. Since they are polychaetes, their bodies consist of multiple segments, each with a pair of parapodia. Their heads have two pairs of eyes at the sides each with lenses. Each pair of eyes is orientated in a different plane. Females can grow up to 20mm while males grow up to about 12mm. Their nickname comes from the fact that they are bioluminescent when they rise to the surface during the mating period. The Bermuda Fireworm typically lives in protected rocky bottoms and they swim to the surface 2 to 3 days after each full moon to spawn. These worms are only bioluminescent during mating times. They also follow a lunar periodicity pattern, which is why the swim to the surface to mate around the time of a full moon. Lunar periodicity is typically seen in response to changes in light intensity from the sun to light intensity from the stars. Typically, the female worm appears first, swimming up from the rocky bottom to make circles at the surface. The females then give off a green glow, looking like "marine fireflies". The glowing, though amusing for human spectators, is designed to attract the attention of males at the bottom. If you look closely, you can see the males rush up to the surface to meet the females. When the males fly up to the surface, they immediately fertilized the released eggs.

"Turks and Caicos Glow Worms 2017" by Alana Olendorf [CC by 4.0]

43 Chapter PHYLUM ARTHROPODA

ARTHROPODA

All members of the Phylum Arthropoda have a distinct, rigid exoskeleton of chitin. Arthropoda are also known as the most taxonomically diverse phylum on the planet. They occupy nearly every known habitat on earth. Estimates state that there are anywhere from 30-100 million different species in this phylum. The segmentation of Arthropod bodies is different from that of Annelids in that they consist of subsections composed of fused segments referred to as tagmatization. Tagmata are the specific sections (i.e. Head, Thorax, Abdomen), which may vary from species to species. Some important features of arthropods are their open circulatory system, molting, coelomate, protostome and the fact that they live in almost all habitats on earth.

Arthropods will molt their exoskeletons, this process of molting is also called ecdysis. Their exoskeletons have three main functions: protection from predators, prevention from desiccation, and locomotion (attachment sites for mussels). Sclerotization is the hardening of the procuticle after molting (also called tanning). Arthropods are also commonly noted for their jointed appendages. Examples of marine Arthropods include lobster, banded coral shrimp, cleaner shrimp. barnacles, horseshoe crabs, copepods, and other microcrustaceans that form the zooplankton.

The most important difference that is noticed between certain Arthropods such as crayfish and crabs is the morphology of and arrangements of body parts. Crabs and crayfish both have chelipeds, but crabs have a distinguishable single structure prosoma.

As with all Arthropods, both types of creatures are contained by chitinous exoskeletons that they are able to shed and regrow throughout their life-cycles.

"Coral Banded Shrimp" by hjk_888 [CC BY-NC-ND 2.0]

"Rode amerikaanse rivierkreeft, Red Swamp Crayfish, Procambarus clarkii 04" by Luc hoogenstein [CC by 4.0]

Barnacles are interesting, even among the variant Phylum of life that is Arthropods. As with all other Arthropods, they are distinguished by a chitinous exoskeleton, however unlike many of their relatives, a barnacle also has an outer cup made of calcium carbonate that resembles a tiny volcano in structure. Barnacles also have internal calcareous plates that can close. This is not only for protection from predators, but also to keep the barnacles from drying out if they are not submerged in the water. There are over 1,400 known species of barnacles, most of which have the notable characteristics of secreting a natural 'glue' that allows them to adhere to the various surfaces that they inhabit. The incredible part is, through scientific observation and testing, the tensile strength of these natural epoxies have been measured at over 5,000 lbs/in^2. The manner in which barnacles feed is through cirri. This is a 'feather-like' appendage that combs through the water for zooplankton and other microorganisms that they can capture and ingest.

"Barnacle Remains under dissection microscope, 2019" by Jason Charbonneau [CC by 4.0]

44 Chapter PHYLUM MOLLUSCA

MOLLUSCS

The phylum Mollusca is defined by several special characteristics. These defining characteristics include a mantle with a mantle cavity, a shell (except where lost), visceral mass, foot, and radula. The odontophore is in the mouth of most mollusks and it supports the radula (a ribbon of teeth). In many molluscs, it moves forward while the radula contacts the food, allowing the mollusc to feed. Mollusca can be found in freshwater, marine and terrestrial habitats. More features of molluscs include bilateral symmetry, soft or unsegmented bodies, respiration via ctenidium, ganglia/nerve comprised nervous system, haemocoel body cavity, etc. [1].

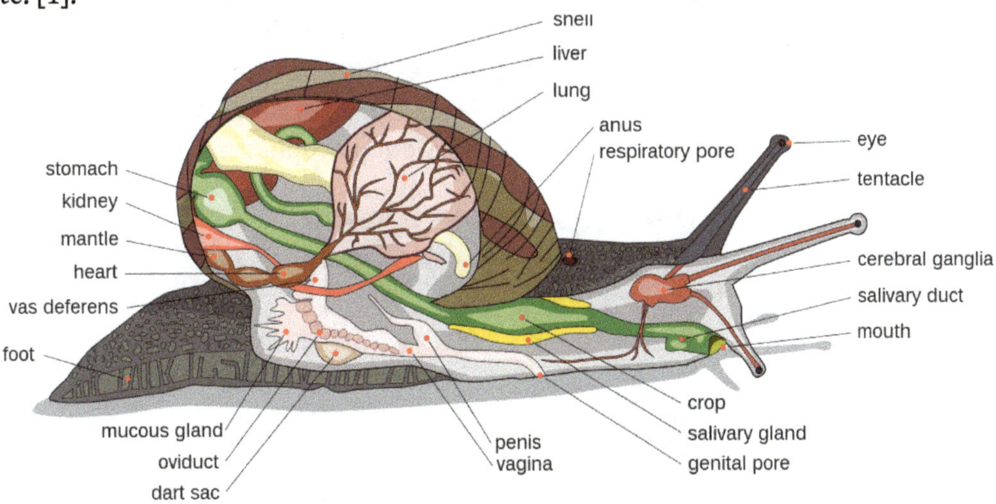

"Anatomy of a Common Land Snail", by Jeff Dahl [CC BY-SA 4.0]

Although when most people think of a mollusc they imagine a typical clam or snail with the shell on the exterior of the organism, there are actually variations of these characteristics. One example of this is the Flamingo Tongue. The Flamingo

Tongue is a small marine snail. It is a fascinating organism, in that their mantle, when they are at rest, covers the outer shell on the exterior of the organism. When threatened the Flamingo Tongue will retract all of the exposed tissue back into the shell in a very interesting manner.

0"Flamingo Tongue, 2018" by Alana Olendorf [CC by 4.0]

POLYPLACOPHORANS

Chitons are the common name for species in the Class Polyplacophora. Chitons are considered more primitive in relation to other groups within the phylum such as bivalves or cephalopods. There are 8 dorsal plates lining the dorsal surface of these organisms. There are many primitive species of the class Polyplacophora that are still extant. However, only a few species of Monoplacophorans still exist in the world today.

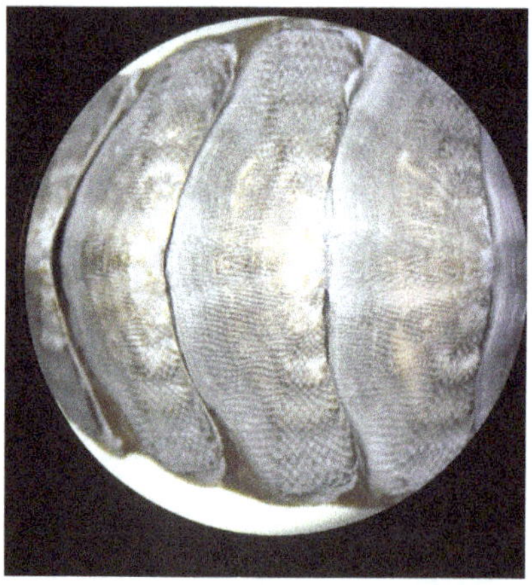

"Dorsal side of chiton specimen under dissection microscope 2019" by Jason Charbonneau [CC by 4.0]

GASTROPODS

Gastropoda is a very diverse class of molluscs. The subclass Prosobranchia, are often identified by their coiled, cone shaped or tubular shells [1]. The mantel cavity is typically located on the anterior of the organism. Important distinctions of these organisms are the variations or the absence of radula. These organisms utilize Nephridia for excretion of nitrogenous waste. One characteristic of gastropods is the presence of a calcareous operculum. This structure acts as a shielded plate that protects the organism by covering the opening when it is retracted into its shell.

In the photo below, the dark oval in the center is the operculum. They are calciferous and rough to the touch, protecting the opening to the shell like a manhole cover.

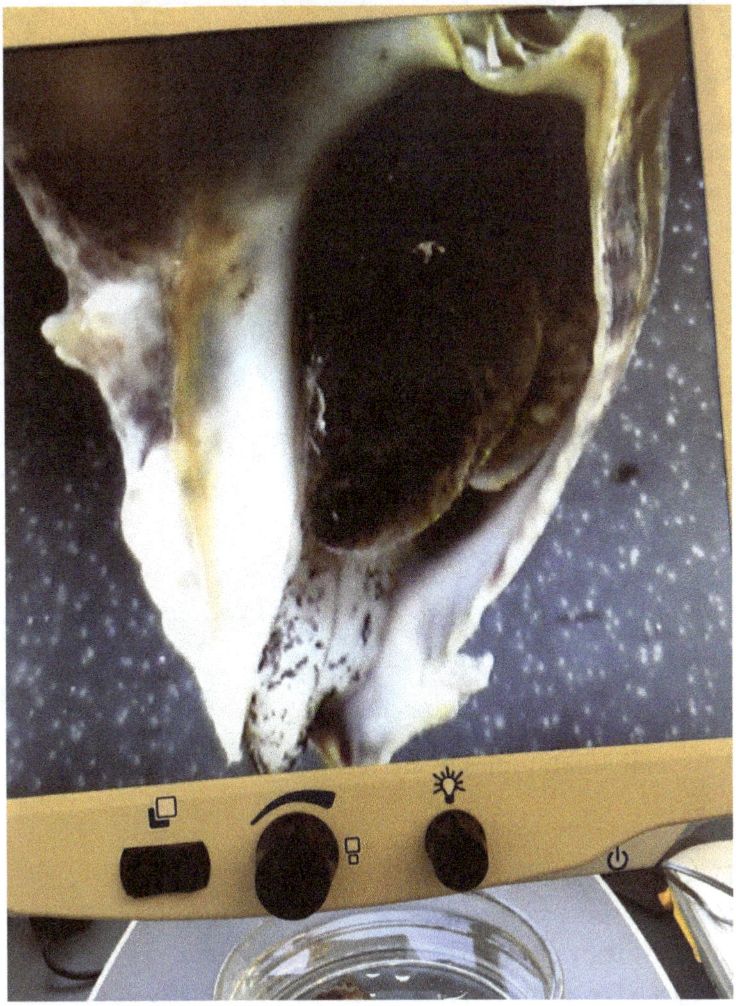

"Gastropod under dissection microscope" by Jason Charbonneau [CC by 4.0]

NUDIBRANCHS

Nudibranchs are typically categorized as sea snails which lack shells. Often they are richly colored and captivating to the eye. Their magnificent beauty is the result of aposematism which is the bright coloration exhibited by these organisms that warns predators that they are toxic, distasteful or dangerous. The toxicity of these organisms differentiates from species to species, with some being exponentially more dangerous than others. Their toxicity is usually contingent upon their evolutionary specialization and their genus's specific niche.

"Flabellina iodinea" by Jerry Kirkhart [CC by 2.0]

BIVALVES

Class Bivalvia consists of molluscs that have two connected shells such as Oysters, Clams, Mussels, Scallops and many more. Most bivalves are enjoyed as delicacies, despite the fact that they are benthic level filter-feeders. As others in the phylum Mollusca, bivalves have a shell that is made up of deposits of Calcium Carbonate. These deposits are derived from substances in the water and harden over-time.

"Clam Shells" by Jason Charbonneau [CC by 4.0]

Visible among the many aspects of an open clam are the posterior and anterior adductor muscles. The function of these muscles is to hold the shell closed as a defense from predators. The strength of these muscles given their relative size is immense. The mantle is a layer of tissue that overlays the visceral mass of these organisms and is directly connected to the shell. The foot of the bivalve is directly responsible for its movement. The muscular foot will emerge when the shell is opened and pushes the organism along or into the benthos or substrate.

CEPHALOPODS

Cephalopods exhibit several similarities but also distinguishable differences from other molluscs. They sometimes exhibit a calcium carbonate shell. Squids and Octopus lack this feature, but the more primitive Nautilus does have this feature. Species in the Class Cephalopoda contain a large closed circulatory system and prehensile arms/tentacles that encompass a mouth, complete with a beak and radula. One of the most important evolutionary aspects of these organisms are their large complex eyes. These eyes are specialized for improved sight at depths where little ambient light reaches. The mantle of these organisms forms a sizable ventral cavity containing ctenidia. A portion of the mantle also forms a muscular funnel. Water is taken up and forced through these chambers under pressure creating a unique form of jet propulsion. There are over 900 living species of cephalopods that inhabit the world today.

"Nautilus shell" by Jason Charbonneau [CC by 4.0]

"Squid, December 9, 2013" [CC 0 Public Domain]

REFERENCES

Brusca, Gary J., Brusca, Richard C. 2003, Invertebrates 2nd ed. ISBN 0-87893-097-3

45 Chapter | PHYLUM ECHINODERMATA

ECHINODERMATA

The Phylum Echinodermata is distinguished by characteristics such as spiny-skin, pentaradial symmetry, and an endoskeleton composed of calcareous ossicles. Given that all species in this phylum are exclusively marine dwellers, they also evolved a specialized water-vascular system. This includes several canals, that comprise part of a hydraulic system for functions such as the extension of limbs, movement, nutrient distribution, gas exchange and feeding. Another fascinating aspect of Echinoderms is that the water vascular system also includes the tube feet with powerful suction capability. Due to the structural makeup of these organisms, largely in part to their calcareous plates, they dry and preserve well as both specimens and fossils.

"Live sea urchin specimen in salt water tank, 2019"
by Jason Charbonneau [CC by 4.0]

ASTEROIDEA

The photo below features a dried sea star. Sea star is a broad ecological term that refers to the entire Class Asteroidea. Sea stars typically have five arms, which they rely upon for movement; these animals also utilize their limbs and the various forms of tube feet on them for the capture and drawing in of prey. Their diet includes but is not limited to: gastropods, bivalves and many annelid worms. Interestingly enough, most Asteroidea in nature protrude their stomach outside of the body through the mouth to digest prey externally. This allows them to digest organisms they have captured and rest on top of. The aboral surface of the specimen shows the spiky-skin described above, jutting out from calcareous plated dermis. The small white structure, slightly off center to the left is the sieve plate or madreporite. This structure is the exterior opening to the water-vascular system of the sea star.

"Aboral view, sea star, 2019" by Jason Charbonneau [CC by 4.0]

The oral surface shows many interesting distinctions. Among them, ambulacral grooves lining the arms. Lining the ambulacral groove are several dozen podia, which are also called tube feet. These podia are the primary mechanism for movement and predation. In the aboral view shown below, the tube feet as the tiny black structures that appear sunken into the organism (this is due to the fact that this specimen is long dead).

"Oral view, sea star, 2019" by Jason Charbonneau [CC by 4.0]

OPHIUROIDEA

The class Ophiuroidea (brittle stars and basket stars) is related to the asteroids but they are more morphologically slender, the arms do not contain tube feet, and they are distally stretched from the central disc. In the case of the basket stars, the structure of their limbs is highly branched, resembling capillary systems, or an entire conglomeration of roots. The complexity of these limbs may be attributed to evolutionary advantages from an increase in the surface area to volume ratio for gas exchange and feeding. The ability to reach out and capture large zooplankton prey without excess expenditure of energy is favorable to these organisms. The brittle star on the other hand exhibits an almost-serpentine like movement of the arms, giving them the nickname serpent stars. The diet of these organisms is very generalist, as they will consume essentially anything that they can come in contact with.

"Brittle Star, 2019" [CC 0 Public Domain]

A special note in the story of the brittle star comes from recent reports that some brittle star species may actually share a symbiotic relationship with corals. An observational study in 2010 showed deep sea corals in the Gulf of Mexico covered in a substance called floc, which is speculated to be an odd combination of trace amounts of oil, dispersants, and excess mucus from stressed or sickly corals. Brittle stars were observed brushing off this substance which would have likely otherwise lead to the death of some very old coral reef colonies [Smithsonian-Ocean]. The extent and mechanism of this relationship is still not fully understood.

ECHINOIDEA

Echinoidea is the class of Echinoderms that includes sea urchins, sand dollars, sea biscuits and others. The spines observed on these organisms are actually mobile, which serves to enhance protection, feeding, and aid in movement. Echinoidea are encased in an endoskeleton commonly called a test. Much like other endoskeleton exhibiting marine dwellers, the test is comprised of calcium carbonate [3]. Other morphological characteristics to note are that sea urchins exhibit longer spines than other members of the class such as heart urchins and sand dollars where the latter are very short relative to the organism's size [4].

“Oral surface of dried sea urchin, 2019” by Jason Charbonneau [CC by 4.0]

“Aboral view of a heart urchin test, 2019” by Jason Charbonneau [CC by 4.0]

HOLOTHUROIDEA

Sea cucumbers are a unique, divergent class from the rest of the Echinodermata. Unlike their counterparts, their 'skeletal' structure is greatly reduced, resulting in a soft body that is malleable and has many captivating traits. Without any arms, class Holothuroidea relies on small tentacles that surround the mouth. They prey on minuscule food items that are afloat in the oceans or that rest on the benthic level of the ocean floor [4].

"Sea cucumber suspended in preservitive, 2019" by Jason Charbonneau [CC by 4.0]

Some unique aspects of sea cucumbers lie in their rather interesting defense mechanisms. If touched or squeezed, they will rapidly shrink in size and project water from several pores all over their body which hardens as they shrink. They can also expel all of their internal organs allowing predators to feed on them, but later they can regenerate these parts which is nothing short of astonishing. They also have mutable connective tissue and can quickly change the shape and texture of their body.

CRINOIDEA

The Class Crinoidea includes the feather stars and sea lilies. The defining characteristic of this class is that they anchor themselves to a substrate through the use of cirri. These cirri are attached to a long stalk which keeps them in place, as most of the species comprising this phylum are sessile. There is an exception to this general rule in the case of *Analcidometra armata*, commonly called the swimming crinoid. This is a rare group of Crinoids that may be observed across the Caribbean; they use their ten arms to gently manipulate water as a medium in order to become mobile [Marine Species Identification Portal]. These Crinoids are threatened species and if seen they should not be touched!

"Feather star, 2006" by Nick Hobgood [CC by 3.0]

REFERENCES

- De Klujiver, M., Gijswijt, G., De Leon, R., & Da Cunda, I. (2019). Swimming crinoid (*Analcidometra armata*). Developed by ETI BioInformatics
- Hall, D. (2018, May 10). A Brittle Star May Be a Coral's Best Friend.
- Hyman, L. H. 1955. *The Invertebrates. Volume IV: Echinodermata.* McGraw-Hill, New York.
- Peachey, Donna & Gordon, The Biocam Museum of Life Series. Kelowna, B.C. Canada VIY 7N8 Box 417 PBC,

Chapter 46 PHYLUM CHORDATA

Four diagnostic features characterize species in the phylum Chordata: 1) The notochord is a malleable rod running the length of the organism's body, to which the rest of the skeletal structure relies upon for foundational support; 2) the presence of a tail extending past the anus; 3) a hollow, dorsal nerve cord (becomes the spinal cord in humans!); and 4) pharyngeal gill slits, with the ability to be modified for specialized functions in mature vertebrates. Though this might sound outlandish, but since we are chordates, even all human fetuses early in development have gills!

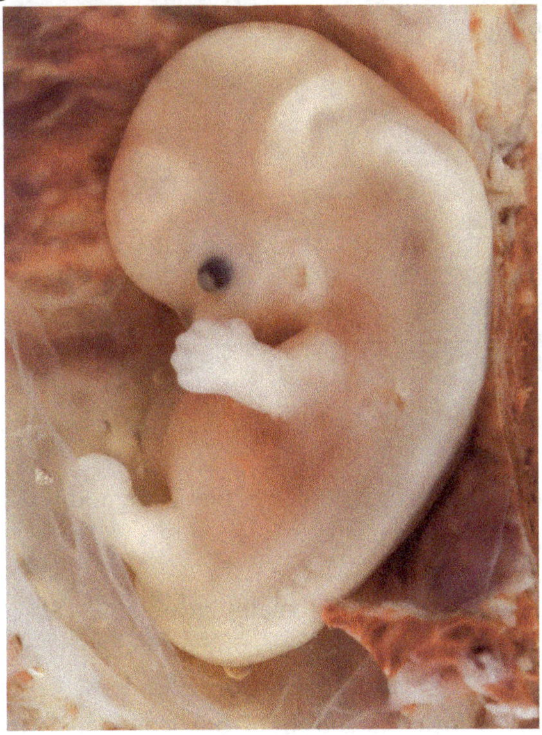

"9-Week Human Embryo, 2000" by Ed Uthman [CC by 2.0]

There are many classes that comprise the phylum Chordata. These classifications and their relationships are constantly reassessed with taxonomic research and our continued development of our understanding of life on Earth. This chapter will focus primarily on the marine chordates.

UROCHORDATA

The subphylum Urochordata includes the tunicates, otherwise known as 'sea squirts'. They are exclusively found in marine environments and it seems strange that they are in the same category as vertebrates. Even more interesting is their manner of nutrient collection and waste expulsion. It is a rather simple system, there is an incoming siphon, that draws in water and food particulates that may be floating in the water. These nutrient particulates are then passed down to the intestine where they are processed for sustenance. The excess water and waste products are expelled through the other siphon known as the excurrent siphon. Even more surprising is the fact that these animals only show all traits of Chordates when they are in larval stage. Urochordata often occur in colonial form as adults as seen below.

"Komodo Tunicate, 2006" by Nick Hobgood [CC by 3.0]

CEPHALOCHORDATA

The subphylum Cephalochordata (lancelets) exhibit all the traits of Chordata as adults. Like Urochordata, they are also marine organisms and may be found world wide in shallow waters. They are often observed in benthic environments, where they burrow themselves into the sediment but leave their anterior exposed as a foraging mechanism [1]. The anterior portion of these organisms resembles that of the face of shrimp or of a praying mantis. Their manner of feeding is through the filtration of nutrient rich waters around them. Cephalochordata have been described as 'fishlike' in comparison to their Urochordata counterparts. The species that comprise this subphylum are relatively small ranging from 5-15 cm in length. And though they possess a closed circulatory system, they have no heart. Instead, their blood is oxygenated via their gill slits and recycled throughout. Their dorsal nerve runs throughout their body, however the anterior end does not form a brain complex.

"Branchiostoma lanceolatum, 1997" by Hans Hillewaert [CC by 4.0]

AGNATHA

Moving into the vertebrate category, it is important to start off with the superclass Agnatha, more commonly known as the jawless fish. Two of the classes comprising Agnatha are Cyclostomata (lampreys) and Myxini (hagfishes). It is commonly accepted that the evolution of vertebrates began with the segmentation of a vertebral column. This gave rise to a 'backbone' [1]. This vertebral column typically incorporates and/or replaces the primitive notochord. Further vertebrate adaptations include the development of sensory organs, a complex neural system, and a brain encased in a skull. Vertebrate Chordates also exhibit bilateral symmetry, having a closed circulatory system with a chambered heart. The degree to which the heart is chambered (i.e. one, two, three and four chambered hearts), varies with taxonomic class.

Lampreys (Cyclostomata) are jawless fish that are parasitic on other fish. As juveniles they derive their nutrition from filter feeding on plankton and particles floating in fresh water. Lamprey juveniles wait until maturity to migrate into salt water environments. In the ocean, lampreys may often be observed attached to larger reef fish and megafauna. They press their mouth to their host and using a tongue, they draw blood and tissue out of their victims.

"Sea Lamprey, 2015" by Joanna Gilkeson [CC by 2.0]

The Myxini Class are commonly known as hagfish and reside solely in marine environments. They are similar in structure to eels and are jawless. Myxini can be either consumers or detritivores, feeding on the flesh of weakened or already dead fish. They have also been known to prey upon small invertebrates. Just like the lamprey, hagfish tongues resemble a rasp, and are similar to serrations on a knife. One of the defense mechanisms exhibited by the Class Myxini resembles that of many amphibians in that when threatened, they will release mass amounts of high-viscosity fluids. These fluids help to distract, escape from, confuse or deter potential predators [1].

"Eptatretus soutii, 2005" by Stan Shebs [CC by 3.0]

OSTEICHTHYES AND CHONDRICHTHYES

Fish can be divided into the two main groups, Osteichthyes and Chondrichthyes. Osteichthyes, traditionally considered as an taxonomic class, is now known to be a paraphyletic group. As their name suggests, they are boney fish that dwell in both fresh water and salt water around the globe. Boney fish are comprised of a hard calcareous skeleton and are coated in slippery, sometimes sharp scales. An important feature of boney fish is their lateral line. This is a zone that runs horizontally along the body of the fish and is predominantly used in the detection of vibrations. The lateral line has been attributed to the coordinated and navigational success of schools of fish, in which mass quantities of individuals conglomerate for various reasons. There are currently over 34,000 known species of fish on the planet, and that number is both growing and shrinking [fishbase]. Many fish populations are threatened by egregious over-fishing practices which has caused species diversity and abundance in various ecosystems to move into a downward spiral.

"1. Oncorhynchus mykiss 2. Pontinus nematophthalmus 3. Aracana aurita 4. Peristedion gracile" by Public Domain [CC 0]

The skeletons of Chondrichthyes are comprised of cartilage rather than bone. Cartilaginous fish include the Rays, Sharks and Chimaeras. The vast majority of sharks are predators, much of their power and deadliness comes from their evolutionary adaptations in their physiology. Their streamlined and highly muscular body lends to their high proficiency as consumers. The flattened bodies of rays contributes to their free-flowing nature through water. Often times rays consume invertebrates that are found in the benthos of the ocean, however they are highly diverse in size, morphology and behavior. Giant Manta Rays can get up to 7m in width and filter feed on masses of zooplankton, whereas the southern stingray may only be a few dozen centimeters in width and feed on the bottom in sandy flats near coral reefs [NOAA].

"Hypanus americanus, 1992" by Barry Peters [CC by 2.0]

CHIMAERA

The Chimaera are an interesting group of organisms that inhabit the deep sea. Their physiology is atypical and they possess a cross of several characteristics as their name suggests. Rather than a jawline filled with teeth Chimaeras have a flat dental plate. Origins of Chimaeroid marine species can be traced back upwards to 280 million years, predating the earliest dinosaurs of the Triassic period.

"Dunkleosteus Skull, 2014" by Zachi Evenor, courtesy of Vienna History Museum, Austria [CC by 3.0]

A January 2017 discovery of a fossilized Dwykaselachus oosthuizeni skull showed that there are few structural differences in ancient Chimaeras compared to modern Chimaeras. CT scans of the fossil showed significant cranial nerves, inner-ear structure, and nostrils, which are all exhibited by modern Chimaeras [UChicago Medicine]. An incredible aspect of this would pertain to Earth's projected geological history. The presence of modern-day Chimaeras with little difference from ancient species means that these organisms survived two mass extinction events, showing the perseverance of the beings that dwell in the ocean depths.

"Modern Deep Sea Chimaera" by NOAA, Public Domain [CC 0]

REFERENCE

Peachey, Donna & Gordon, The Biocam Museum of Life Series. Kelowna, B.C. Canada VIY 7N8 Box 417 PBC, 2000

INDEX

www.ingramcontent.com/pod-product-compliance
Lightning Source LLC
LaVergne TN
LVHW062238200926
848484LV00006B/1566